Routledge Introductions to Development
*Series Editors*
John Bale and David Drakakis-Smith

T0199795

# World Hunger

Eight hundred million people suffer from hunger. Images of the hungry are often in our papers but so too are images of opulence and obesity. Can one framework help us understand these contrasts? This text exposes the links, from the international to the familial, that help explain the diversity of diets in the contemporary world; to explain why hunger persists while agricultural production increases and genetic engineering revolutionises food production and distribution.

Examining the processes which govern access to food in the global supermarket, the book explores the nature and extent of contemporary world hunger and proceeds to analyse specific global, national, sub-national and familial processes which explain why some eat rich, diverse diets and others do not. Using extensive case material to illustrate the connections between the global and local, *World Hunger* asserts that the contrasting material realities of North and South are very much part of the same picture. The author goes on to identify some positive changes, discussing places and spaces within which people challenge prevailing material inequalities and the ideologies which perpetuate them in an effort to provide workable solutions to the world's hunger problem.

**Liz Young** is Senior Lecturer in Geography at Staffordshire University.

# In the same series

John Cole
*Development and Underdevelopment*
*A Profile of the Third World*
Janet Henshall Momsen
*Women and Development in the Third World*
David Drakakis-Smith
*The Third World City*
Allan and Anne Findlay
*Population and Development in the Third World*
Avijit Gupta
*Ecology and Development in the Third World*
John Lea
*Tourism and Development in the Third World*
John Soussan
*Primary Resources and Energy in the Third World*
Chris Dixon
*Rural Development in the Third World*
Alan Gilbert
*Latin America*
David Drakakis-Smith
*Pacific Asia*
Rajesh Chandra
*Industrialization and Development in the Third World*
Mike Parnwell
*Population Movements and the Third World*
Tony Binns
*Tropical Africa*
Jennifer A. Elliott
*An Introduction to Sustainable Development*
*The Developing World*
George Cho
*Global Interdependence*
*Trade, Aid, and Technology Transfer*
Ronan Paddison
*Retail Patterns in the Third World*
David Simon
*Transport and Development*

Liz Young

# World Hunger

London and New York

First published 1997
by Routledge
11 New Fetter Lane, London EC4P 4EE

Transferred to Digital Printing 2003

Simultaneously published in the USA and Canada
by Routledge
29 West 35th Street, New York, NY 10001

Typeset in Times by
Pointing–Green Publishing Services, Chesham, Buckinghamshire

*British Library Cataloguing in Publication Data*
A catalogue record for this book is available from the British Library

*Library of Congress Cataloging in Publication Data*
Young, Liz
    World hunger / Liz Young.
    (Routledge introductions to development)
    Includes bibliographical references and index.
    1. Food supply – Developing countries. 2. Hunger – Developing
countries.   I. Title.   II. Series.
TX360.5.Y68        1996
363.8'09172'4–dc20        96–36545

ISBN 0–415–13773–X

For Jim

# Contents

List of plates                                                              ix
List of figures                                                             xi
List of boxes                                                               xii
List of tables                                                              xiii
Acknowledgements                                                            xiv

1  A framework for analysis and historical overview                         2
   Introduction                                                             2
   A framework for analysis                                                 3
   Historical perspectives                                                  7
   Conclusion                                                               14
   Key ideas                                                                15

2  The contemporary nature and extent of hunger                            17
   Terminology                                                              17
   The contemporary extent of hunger                                        26
   Conclusion                                                               30
   Key ideas                                                                34

3  International perspectives on global hunger                             36
   Introduction                                                            36
   The changing geography of global food production and
   consumption, from the sixteenth to the twentieth century                37
   Trends since the 1970s: globalisation and transnational
   corporations (TNCs)                                                      47
   Conclusion                                                              59
   Key ideas                                                               59

**4   National perspectives**                                      **64**
Introduction                                                        64
Historical legacies                                                 66
Development strategies                                              70
Conclusion                                                          83
Key ideas                                                           85

**5   Gendered fields**                                            **88**
Introduction                                                        88
Women as economic actors                                            91
Intra-household entitlements                                        99
Conclusion                                                         106
Key ideas                                                          109

**6   Sub-national perspectives**                                  **111**
Introduction                                                       111
Regions                                                            111
Ethnic and religious minorities                                    120
Households and food security                                       121
Conclusion                                                         130
Key ideas                                                          131

**7   Conflict and hunger**                                        **133**
Introduction                                                       133
Conflict and hunger: the connections                               134
Conclusion                                                         144
Key ideas                                                          145

**8   Alternative futures**                                        **147**
Introduction                                                       147
Whose business is it?                                              148
Empowerment approaches                                             161
Conclusion: concerted actions                                      166
Key ideas                                                          168

**Review questions, reference and further reading**               **169**
**Index**                                                         **177**

# Plates

3.1  Exotic foods                                                    50
3.2  We're all global now                                            51
3.3  What sort of apple did that recipe recommend?                   51
3.4  Agricultural machinery, Kansas, USA                             56
3.5  Hybrid maize varieties, Kansas, USA                             56
3.6  The Hoover Dam, Colorado, USA                                   57
3.7  Local millet porridge, northern Nigeria                         60
3.8  An Islamic butcher, Kano, Nigeria                               60
3.9  Fishing in the River Sokoto, Nigeria                            61
3.10 A sugar cane seller, northern Nigeria                           61
4.1  Female labourers in India                                       76
4.2  Seamstresses in Kano city, northern Nigeria                     76
4.3  Beating indigo dye, northern Nigeria                            77
4.4  Fuelwood for the urban market, northern Nigeria                 77
4.5  Urban street trader, Calcutta, India                            80
4.6  Street hawker, Dhaka, Bangladesh                                80
4.7  Urban market, Dhaka, Bangladesh                                 80
4.8  Urban agriculture, Dhaka, Bangladesh                            81
4.9  Urban agriculture, Delhi, India                                 81
4.10 Recent migrants to Delhi                                        82
5.1  Cultivating a *dambo* in Zimbabwe                               92
5.2  Maize farmers in Zimbabwe                                       92
5.3  Women harvesting olives in Tunisia                              93
5.4  Woman winnowing guinea corn in Kano state, Nigeria             93
5.5  Women harvesting mussels from wetlands outside Calcutta        100

5.6   Women selling craft work to tourists, small town,
        Ecuadorian Andes                                            100
5.7   Women selling *kosai* in northern Nigeria                     101
5.8   Female vegetable trader, Jos Plateau, Nigeria                 101
6.1   A farmer in Zimbabwe battles against soil erosion            116
6.2   Soil erosion in a semi-arid region in Tunisia                116
6.3   Animal market, Sokoto, Nigeria                               119
6.4   Fulani women transporting milk to market                     119
6.5   Guava tree, Kano state, northern Nigeria                     128
6.6   A village on the Jos Plateau, Nigeria                        128
6.7   Kebili Oasis, Tunisia                                        128

# Figures

2.1 The vicious circle of female malnutrition 22
2.2 Distribution of sample countries by per capita daily
    calorie intake 28
4.1 How governments interact with the food security system 65
6.1 India: regional diversity 113
6.2 China: regional contrasts in rural income and literacy 122
6.3 The local food security system 124
8.1 World per capita cereal production, 1951–93 151
8.2 Diversity of NGO types 162

# Boxes

1.1 The quality of food and drink in nineteenth-century England    8
1.2 Cholera in Manchester, 1832    13
2.1 Women and anaemia in the developing world    21
2.2 The breastfeeding debate    23
2.3 The northeast of Brazil    31
3.1 Wheat and the pampas    38
3.2 Get fresh!    48
3.3 Industrial agriculture and the environment    55
4.1 China: a conflicting picture    67
4.2 Cash crops and entitlements: some dangers    73
4.3 Harare: the urban food system    79
5.1 Village women pack up and leave    94
5.2 Joyce Kayaya, a Zambian farmer    96
5.3 Contrasting cases: gender relations and household
     expenditure    103
5.4 Female empowerment    107
6.1 Mali    118
6.2 Sri Lankan case study    127
7.1 Refugees and displaced people    139
7.2 Children sold by Afghan widows    142
8.1 Trends in world food production    151
8.2 World Summit for Children/International Conference on
     Nutrition    155
8.3 Brazil's campaign against hunger and misery, and for life    163

# Tables

| | | |
|---|---|---:|
| 1.1 | The construction of entitlements | 5 |
| 2.1 | The percentage of chronically underfed, by region | 27 |
| 2.2 | The absolute number of chronically underfed, by region | 30 |
| 2.3 | GNP per capita and selected social indicators, selected countries, 1992 | 31 |
| 3.1 | The power of the transnationals | 54 |
| 4.1 | Growing urbanisation in selected countries, 1992 | 71 |
| 5.1 | HDI and GNP rankings, 1992 | 90 |
| 5.2 | HDI and GDI values, selected countries, 1992 | 91 |
| 5.3 | Changes in average GDI values, 1970–92 | 106 |
| 6.1 | India, 1992: urban–rural discrepancies in access to basic services | 112 |
| 6.2 | Regional diversity in India | 114 |

# Acknowledgements

Several colleagues, past and present, in the Geography Division at Staffordshire University were very helpful. I would like to thank Jenny Elliott and Steve Wyn Williams for their initial encouragement, members of the Research Committee who granted me time off from teaching, and Sabiha Sayid, Hamish Main and Jenny Elliott for being so generous with their slides and time. Map work was completed with the help of Owen Pucker and John Mooney who were always quick and efficient, thank you both. Finally, I would like to thank David Drakakis-Smith and John Bale for their help and encouragement throughout.

# Vietnam

## SHRIMP AND CHICKEN SALAD                                     Serves 4

Each chapter opens with a recipe. These are included for several reasons: to indicate the richness and variety of food around the world which exists despite recent global processes of homogenisation; to encourage readers to experiment with different recipes and consider how these differ from their diets; to remind readers that all over the world, even when people are poor, families and friends gather to eat and enjoy food. In addition to its function in maintaining our health, the social and cultural significance of food is immeasurable.

## INGREDIENTS
½ lb/225g shelled shrimps, cooked
6-inch/15cm cucumber, sliced
handful fresh cilantro/coriander leaves
½ lb/225g cooked chopped lean chicken meat, cut into small cubes
½ cup/60g peanuts, crushed or coarsely chopped
4 cups/200g bean sprouts
salt and pepper
2 tablespoons lemon juice
1–2 teaspoons sugar
2 tablespoons fish sauce*
1 carrot, grated

* Fish sauce, called *nuoc nam*, is obtainable from Chinese food shops; or you could use 1 tablespoon anchovy essence instead.

## METHOD
1  Pour in just enough water to cover the bottom of the saucepan. Heat it up and when it is boiling throw in the bean sprouts, for a few seconds only. Drain them and allow to cool.
2  In a bowl, mix the lemon juice with the sugar, fish sauce, salt and pepper.
3  Now arrange the bean sprouts, carrot and cucumber on a serving dish and pour the dressing over.
4  Place the chicken and shrimps on top of the salad and garnish with the cilantro/coriander leaves and crushed peanuts.

# 1
# A framework for analysis and historical overview

## Introduction

> Sainsbury's marketing manager described entering the store as a geography lesson or a trip around the world.
>
> (Cook, 1994, 244)

This book examines the geography of the world food system. It examines the processes 'behind the supermarket shelves' which explain the geography of food production and consumption. The main thesis is that hunger persists because the political will to eliminate it is lacking. Decisions made at all scales, from the international to the familial, help explain why some people enjoy a rich and varied diet while others suffer from hunger. This book challenges traditional conceptualisations of hunger, which analyse it with reference to natural disasters and 'overpopulation' and which tend to grant it an element of inevitability. There is nothing inevitable about the persistence of hunger. When the essential political character of hunger is appreciated then it becomes possible to envisage a world where hunger is history.

While the political character of the problem has long been appreciated by some academics (Warnock, 1987), the 'problem of hunger' in popular consciousness and in some textbooks continues to assume an apolitical character which denies the connections between feast in some regions and hunger in others. It is conceptualised as a 'world food problem' rather than a problem of 'world hunger'; these are quite different things. Most students, when asked to rank the causes of world hunger, prioritise natural causes over human ones; floods, droughts and poor soils are most popular. When the human dimension is acknowledged, the 'problem of population' is most frequently offered, followed by war. Several other assumptions are exposed through discussions with students. Among the most relevant are the following:

- that hunger exists only in the developing world;
- that hunger in the developing world is explicable with reference to the internal characteristics of those countries alone, that is they are ignorant of the historical and international dimensions of the problem;
- that famine is the main problem;
- that the problem of hunger is most serious on the African continent;
- that increased food production is imperative.

This text challenges all of these easy assumptions.

## A framework for analysis

Many students subscribe to the 'lack of' school of thought, usually associated with the 'cycles of poverty' school. This theory is based on the notion that where hunger persists it is because people lack everything from 'good weather' to 'modern technology', 'the pill' and 'education and investment', and that all these reinforce each other. These mantras prove incredibly resilient and have been known to emerge in examinations at the end of a series of lectures and seminars specifically designed to undermine them as explanations of hunger. This text reflects my efforts, in lectures and seminars, to alter these widely held assumptions so that we, individual consumers in the affluent world, can appreciate how we are implicated in the problem of world hunger.

Since the middle of the nineteenth century the most popular assumption has been that hunger is caused by population growth, that population growth proceeds at a faster pace than food production and that famine occurs if population growth is not reduced. This is testament to the persistence of Malthusian interpretations of hunger, which are adequately critiqued elsewhere (Arnold, 1988; Devereux, 1993; Lappe and Collins, 1986) and need not be repeated here. Recent analyses have turned Malthus on his head, so to speak, by understanding that poverty *causes* population growth rather than vice versa. The best way to reduce fertility is to raise the living standards of the poor and to improve the status of women in society and their access to power, which may include increasing their ability to make informed decisions about their fertility (Momsen, 1991). It has also been observed that population can expand rapidly and not be accompanied by widespread hunger – China since 1949, except for the 1958 famine – or indeed that food production per capita can increase without being accompanied by any diminution of the problems of hunger. Clearly, a more sophisticated framework is required to reveal the complexities of global hunger.

## Proximate and structural causes of hunger

A useful first step is to differentiate between proximate and structural causes of hunger. Proximate causes of famine and/or undernutrition are those which can be identified immediately as playing a role. Some of the most important are war, drought, flooding, late rains, and crop failures due to disease or pests. A recent example is the chaos to food supplies in North Korea because of floods in 1995 and torrential rains in 1996. The food crisis that ensued caused the World Food Programme to launch an emergency aid operation in 1995 which was expanded in 1996; it is currently supplying food aid to approximately 1.5 million people in North Korea. This analysis occasionally considers the proximate variables listed above, but prioritises instead the role of long-term structural processes and the political context of hunger creation. The thesis is that while proximate variables trigger hunger or famine, these are only effective as triggers in specific 'spaces of vulnerability' (Watts and Bohle, 1993) that have emerged consequent upon historically created processes and ideologies which dictate access to power, in its many manifestations, at the international, national and local levels.

## The entitlement concept

> Starvation is the characteristic of some people not having enough food to eat. It is not the characteristic of there being not enough food to eat. While the latter can be the cause of the former, it is but one of the many possible causes.
>
> (Sen, 1981, 1)

Since the publication of Sen's (1981) research on hunger a new concept has been available to inform analyses of hunger. That is the concept of entitlement. Some of the limitations of the term as initially outlined by Sen have been addressed and an elaborated notion of entitlement has been constructed which is more comprehensive. It is this formulation that informs the analysis outlined in this text. The analytical focus is upon understanding distributive mechanisms; that is, what determines how available food is distributed and how politics, economics and ideology influence distribution. A main proposition is that concepts used to analyse famine are as relevant to analyses of chronic hunger, of which acute hunger, manifest as famine, is a part. The geography of hunger is explained by employing an elaborated conceptualisation of entitlement and examining how it controls command over food at a variety of scales. Table 1.1 shows some of the factors and mechanisms at every scale which influence peoples ability to command food. The following discussion summarises the main elements of the term and indicates its utility in exposing the social relations that explain the gross inequalities in access to a decent diet which exist between populations.

**Table 1.1** The construction of entitlements: selected factors and mechanisms at different scales

| | Factors/Mechanisms | | |
|---|---|---|---|
| Level of analysis | Historical | Economic | Political/ ideological |
| International | processes of integration | BWI (WB, IMF and GATT/WTO)* | development philosophies |
| | accumulation of power and wealth in core | TNCs** gendered macro-economic policy | SAPs*** interpretations of hunger |
| National | processes of integration character and capacity of the state property relations access to resources | export orientation investment priorities, infrastructure, health, etc. | development strategies gender relations access to power public policy |
| Regional | processes of integration regional and ethnic disparities environmental legacies | limited infrastructural development vulnerable income base | public policy regional/ethnic discrimination differential representation in government |
| Household | socially sanctioned familial patterns | investment decisions | gender relations public policy access to power and resources |

*Note*: Neither factors nor scales are discrete and factors and mechanisms are indicative only.
* BWI = Bretton Woods Institutions; WB = World Bank; IMF = International Monetary Fund; GATT = General Agreement on Tariffs and Trade; WTO = World Trade Organisation.
** TNC = transnational corporations.
*** SAP = structural adjustment plan.

The patterns of food distribution, at a variety of levels, may be examined with reference to people's entitlements, reflected in their ability to *command* food. This term, 'entitlement', used by Sen, may be employed more generally than he suggested by understanding historical and contemporary patterns of food production and distribution as reflecting the relative power of people in different times and places to command food. The ability of some people to command (acquire) food may reflect their political, military, economic or inherited position within the international system and its national and sub-national elements. The term 'command' is used because it suggests that an individual's or group's ability to acquire food is correlated with their access to power, however expressed and at whatever level. Elements which determine entitlements, and hence people's command over food, are now explored at the international, national and local levels.

As employed here the term 'entitlement' refers to the power of any individual or group to acquire a decent diet: the ability to command food. In Britain, any person's entitlement comprises capital they may possess, income they earn from selling their labour, supplements from the state and assistance from family. In rural areas of the developing world many peasants have entitlements based on a small plot of land, from which they may produce food for consumption or sale, occasional earnings from selling their labour, earnings from the sale of domestic production and assistance from other family members. Urban dwellers in the South often depend on very insecure entitlements: earnings from casual labour, crops grown on common land, publicly subsidised food programmes, etc. Generally, in the developing world there is no generalised state assistance, although in some areas there may be an expectation that, in a crisis, help is provided by the state or as charity.

Commonly the term 'entitlement' is used to explain how specific groups within any society command food. Groups of people can be identified who have similar entitlement packages at any given time. In contemporary Britain the homeless have vulnerable and limited entitlement packages. In the developing world landless rural people have limited and precarious entitlement provision and urban, unemployed people in the cities in the South have varied and insecure entitlements. Entitlement packages are dynamic and alter as social change differentiates groups by class, ethnicity, age, region and gender, and as those with limited entitlements try to extend them by a variety of methods – squatting on land, colonising land, fighting for changes in public policy, etc.

The term 'entitlement' may be used to understand differential power relations in international relations too. The United States and the countries in the European Union (EU) have diverse and generous aggregate entitlement packages based on their natural resource endowment and their inherited status as major world powers. This latter status grants them greater influence in modifying and directing the contemporary world economy. Many of the states in sub-Saharan Africa are at the less privileged extreme on the international entitlement continuum; they have weak economies and little political influence to exert for the advancement of their national ambitions. Countries classified as newly industrialising (NICs) fall in between these two extremes and are extending and diversifying their economies and political influence in international affairs. Gross domestic product per capita (GDP pc) and human development indices (HDI) rankings are a rough guide to any country's entitlements, although it is important to stress that as with all aggregate statistics these conceal great differences within countries.

## Historical perspectives

A review of mortality decline in Western Europe is valuable because it establishes the crucial role of social, political and institutional changes in the disappearance of famine and widespread hunger. Demographic transformations in Western Europe during the late eighteenth century and nineteenth century have been the focus of sophisticated technical and theoretical attention since the 1960s. Most important for our purposes are the analysis of spatial and temporal variations in mortality rates. Recent interpretations of the mortality decline in Western Europe can be understood with reference to changes in people's entitlements. The evidence is briefly considered below.

It is difficult to appreciate how drastically our lives in Western Europe at the end of the twentieth century contrast with those of Europeans in previous centuries, even the last century. One of the most dramatic manifestations of that contrast is mortality rates. In previous centuries people lost family and friends more rapidly than we do today; consider the following average demographic experience of the eighteenth century:

> Of every 1,000 infants, only 200 would go on to the age of 50, and only 100 to the age of 70. A man who had beaten the odds and reached his half-century would, we imagine, have seen both his parents die, have buried half his children and, like as not, his wife as well, together with numerous uncles, aunts, cousins, nephews, nieces, and friends. If he got to seventy, he would have no relations and friends of his own generation left to share his memories.
>
> (McManners, 1981, 50)

## The Great Irish Famine

Famines, or food crises, occurred in Europe until the nineteenth century, and the majority of people were poorly nourished; diets for all but the privileged few were boring and inadequate, and until the last decade of the nineteenth century food and drink were liable to be adulterated (Box 1.1). The last famine in Western Europe accompanied by massive increases in mortality was the Great Irish Famine, which was at its most serious during the winter of 1846/47. The Irish population in 1841 was approximately 8.2 million. After the potato failures in 1845, 1846 and 1847 and the deaths and migrations which followed in its wake, the population in 1901 was approximately 4.5 million. During the famine over a million men, women and children died. The proximate cause of the food crisis was a fungus, which devastated the potato crop that formed the bulk of the diet of the poor. The structural causes were more complex and are more contentious but certainly a 'space for famine' had been created. That space was the result of the particular process of capitalist incorporation which had been integrating Ireland into the British and larger North Atlantic economy since the seventeenth century, the specific changes in capitalism that had occurred in the decades which preceded the Famine and finally, but perhaps most crucial, the prevalence of ideologies about the nature of poverty and the source of funds to relieve it.

**Box 1.1**

---

### The quality of food and drink in nineteenth-century England

Until the 1870s bread production in London and the large towns was dispersed among a multitude of small, competitive, primitive bakeries. The typical bakehouse oven was built in a cellar under the roadway. The mixing troughs and kneading boards were in an uncleaned, vermin-infested basement. Bakers worked through the night and it was normal to lock them in to prevent stealing, or drinking, while unsupervised. The usual temperature in the basement was 80 to 110 degrees Fahrenheit. Some bakehouses had a privy under the stairs, but in the even less capitalised concerns the men relieved themselves on the coal heap. The men used both hands and feet while kneading the dough, sweating as they worked. They washed in the water used for the next batch of dough.

As control of adulteration strengthened, tampering contracted to the two mass-consumption perishables, milk and butter. The milk started in bad surroundings and got worse. Until the 1860s in London, and elsewhere until the twentieth century, milking cows were kept crowded in yards, cellars or closed sheds within cities and towns. The standard feed was brewers' grains and distillers' wash. This gave the milk a distinctive taste and the cowsheds a distinctive 'offensive smell'.

---

**Box 1.1 (*continued*)**

Dairymen believed that the more immobilised, by crowding, the cow was kept, the less food she consumed and the more milk she gave. Until 1862 in London and 1879 elsewhere, there was no law requiring cowsheds to be regularly cleaned. Dairies were run jointly as slaughterhouses. Disease was rampant. Beasts *in extremis* were quickly dispatched as meat, so disease could not be calculated.

In 1902 butcher Harris of Clerkenwell suffered his second conviction for selling bad meat. He had pork that was 'suppurating and decompose', and veal that was 'green and slimy'. Harris claimed that the meat was only 'muggy' and would be 'alright [when] it was wiped'. Beef always became cheaper during outbreaks of disease. In mid-1861 farmers were getting 2d a pound for the meat of diseased beef and sheep.

One clear lesson is that, among the working classes and the poor, all [food] choices were dirty ones. Take meat. The crucial point is not how much meat the lower classes got but its quality. Meat at the ruling prices was sold in at least four grades: first to third, and then 'inferior'. But below 'inferior' there was an enormous trade in cheap offal and old and diseased meat. In the countryside, before the 1850s, bullocks' heads and ox-cheeks were never seen; a sheep's head had to be 'bespoken weeks' before the sheep was killed and sheep's pluck 'cost too much' for agricultural labourers' families. Within the family, both rural and urban, into the mid-1860s, wives who were not delicate or who did not go out to work were said 'never' to eat meat. Throughout the century there was no effective control on the quality of meat at the point of slaughtering. Street vendors of food, unlike hawkers, remained unlicensed until at least 1912.

*Source*: Smith (1979).

The capitalist incorporation of Ireland between the sixteenth and nineteenth centuries was aided by English and Scottish colonial plantations, which had created a system of property relations that were hierarchical and unequal. There were a few privileged landowners, who owned estates from the vast (over 10,000 acres) to the more modest (1,000–10,000 acres), but the great bulk of the population had farms of less than thirty acres or were squatters who had colonised common land or had no rights to land at all. While most of the largest landowners were from the settler populations, not all were, and certainly a great many of the settlers were not privileged but small tenant farmers too. Between the late eighteenth century and the 1840s changes in economic circumstances meant that unemployment and underemployment intensified in Ireland. The rurally based proto-industrial (cottage) industries, especially the textile related, which had been so vital in supplementing household incomes, collapsed in the face of competition from industrial producers in Belfast, Derry and industrial cities in Britain. The 1840s was a decade of economic hardship all over Western Europe (there were food riots in England and serious food crises across Europe), so income from temporary migration to Britain also declined in this decade.

On the eve of the Famine at least 40 per cent of the population had very limited entitlements and lived on a diet based on the potato. When that crop failed, three years in succession, these populations were at the mercy of the public purse and charitable donations; neither source proved sufficient to alleviate the impact of the crop failure. It is increasingly realised that in this case, as in other famine conditions, it was social dislocation allied to disease that caused the high death rates, not starvation. Typhus, typhoid and cholera were all present between 1846 and 1850 and killed the poor and some of the better fed too (for detailed accounts see Young, 1996b; Daly, 1986).

## Hunger and disease: some connections

Famines or food crises did not always result in increases in mortality rates but, as in parts of the developing world today, they resulted in social dislocation, migration and disease. With the exception of the Irish example, however, famines had ceased to be a major cause of high mortality in eighteenth- and nineteenth-century Europe. Instead, high mortality rates were the result of the prevalence of poor nutrition in an environment where diseases were rampant. The links between malnutrition and disease are complex and have physiological as well as social elements. First the physiological connection:

> Particular diseases were the indispensable infantry in Death's dark armies, but his generals were Cold and Hunger.
>
> (McManners, 1981, 41)

Here again is an example of proximate and structural causes. Diseases were usually noted as the cause of death in the seventeenth, eighteenth and nineteenth centuries; these were the proximate cause of death. To appreciate why so many were prone to infection, however, it is necessary to consider the cold, unhygienic and hungry circumstances which the majority of the population suffered periodically. Death rates peaked in the hungry and cold seasons of autumn and spring because at these times populations were more vulnerable to infection. As today, different groups within any country experienced different mortality rates. Although epidemics, once established, are no respecters of class, gender, age or ethnicity, many of the chronic diseases were more prevalent in the poorer populations, who suffered from malnutrition and lived in miserable conditions.

Social disruption associated with food shortage forms a second link between food shortages and disease. When food crises occur the weak and hungry migrate in search of food. These populations often accumulate in unsanitary conditions, where infectious diseases flourish. The most serious diseases associated with these circumstances are typhus, known as 'famine fever', typhoid, cholera, smallpox, pneumonia and the plague, which the French historian Le Roy Ladurie called the 'holocaust of the undernourished'. The role of cholera as a cause of death in past and present famines is well documented:

In India repeatedly throughout the nineteenth century epidemics of cholera multiplied famine mortality two- or three-fold; in the Russian famine of 1891–92 nearly half of the 650,000 deaths reported were attributed to cholera.

(Arnold, 1988, 24)

In the past, as in the present, malnutrition and disease have a complex and symbiotic relationship. Malnourished populations, as outlined above, are more prone to infectious diseases but many illnesses also precipitate malnutrition. Any of the intestinal illnesses prevent the body from exploiting the food consumed and thereby intensify the problem of malnutrition. When the poor eat poor-quality food they contract intestinal problems, which reduce the ability of their bodies to exploit the food they eat. Circumstances in Europe as recently as 100 years ago were grim for the many, but circumstances changed and mortality rates fell during the nineteenth century. Why?

## The decline of mortality in Western Europe

Traditional interpretations of the mortality decline in Britain stressed the role of technological advances. Technology was deemed important in two ways. Increases in food production were explained by improvements in agricultural technology associated with the agricultural revolution of the seventeenth and eighteenth centuries. In addition, because of advances in transport technology, food was imported from new source areas increasingly further afield. More food was available, and although the population expanded rapidly between 1750 and 1900, it was sustained by domestic increases in food production and access to new sources of food. Advances in medical technology were also granted a special role in traditional explanations of mortality decline in the nineteenth century.

Recent interpretations of the mortality decline have emphasised other factors, however. Chief among these is the role of the state in provisioning its populace when famine threatened. Arnold argues that:

> in the European context the need to mitigate famine and provision the people was one of the most important factors behind the rise of the modern nation-state, just as neglect of this responsibility exposed regimes to some of their most serious challenges.
>
> (Arnold, 1988, 104)

Increasingly, technological advances offered the opportunity for those in power to intervene to alleviate hardship and to provision expanding urban populations but the crucial variable was the *political will* to intervene. As Europe's emerging nation-states became territorially integrated and expanded their military and administrative capacities, legislative and institutional changes evolved to ensure that food supplies were secured. Mortality rates declined in Europe because it became politically rational for the political élite to intervene to relieve famine and the spread of diseases. Reductions in disease certainly owed a great deal to the political decision to invest public money in public health-related schemes. The role of medicine was important only because medically trained people agitated for public health reforms and, because of their status, their pleadings resulted ultimately in policy innovations, not because their actual medical interventions were impressive: serious successful medical interventions occurred only after the initial fall in mortality rates (Box 1.2).

Mortality rates, like malnutrition statistics, vary by age, gender and ethnic group, employment, region, and between rural and urban areas. Throughout most of the nineteenth century mortality rates were higher in urban areas because of poor sanitation facilities, were higher for infants than adults, were higher for immigrant groups, and were higher than average for females in their reproductive years. Likewise, mortality decline was differentiated by all these variables.

Box 1.2

**Cholera in Manchester, 1832**

**Sir James Kay-Shuttleworth**

I had requested the younger members of the staff, charged with the visitation of the outpatients of the infirmary, to give me the earliest information of the occurrence of any cases indicating the approach of cholera. I had a scientific wish to trace the mode of its propagation, and to ascertain if possible by what means it would be introduced into the town. My purpose also was to discover whether there was any, and if so what, link or connection between the physical and social evils, to which my attention had been so long directed.

A loop of the River Medlock swept round by a group of houses lying immediately below Oxford Road, and almost on the level of the black, polluted stream. This was a colony of Irish labourers and consequently known as Irishtown. I was requested by one of the staff of the outpatients of the infirmary to visit a peculiar case in one of these cottages. He gave me no description of it as we walked thither. On my arrival in a two-roomed house, I found an Irishman lying on a bed close to the window. The temperature of his skin was somewhat lower than usual, the pulse was weak and quick. He complained of no pain. The face was rather pale, and the man much dejected. None of the characteristic symptoms of cholera had occurred, but his attendant told me that the strength had gradually declined during the day, and that, seeing no cause for it, he had formed a suspicion of contagion. I sat by the man's bed for an hour, during which the pulse became gradually weaker. In a second hour it was almost extinct, and it became apparent that the patient would die. His wife and three children were in the room, and she was prepared by us for the too probable event. Thus the afternoon slowly passed away, and as evening approached I sent the young surgeon to have in readiness the cholera van not far away. We were surrounded by an excitable Irish population, and it was obviously desirable to remove the body as soon as possible, and then the family, and to lock up the house before any alarm was given. As twilight came on the sufferer expired without cramp or any other characteristic symptom. The wife had been soothed and she readily consented to be removed with her children to the hospital. Then

Box 1.2 (*continued*)

suddenly the van drew up at the door, and in one minute, before the Irish were aware, drove away with its sad burden.

No case of Asiatic cholera had occurred in Manchester, yet notwithstanding the total absence of characteristic symptoms in this case, I was convinced that the contagion had arrived, and the patient had been its victim. The Knott Hill Hospital was a cotton factory stripped of its machinery, and furnished with iron bedsteads and bedding on every floor. On my arrival here I found the widow and her three children with a nurse grouped round a fire at one end of a gloomy ward. I ascertained that all necessary arrangements had been made for their comfort. They had an evening meal; the children were put to bed near the fire, except the infant, which I left lying upon its mother's lap. None of them showed any sign of disease, and I left the ward to take some refreshment. On my return, on a later visit before midnight, the infant had been sick in its mother's lap, had made a faint cry and had died. The mother was naturally full of terror and distress, for the child had had no medicine, had been fed only from its mother's breast, and, consequently, she could have no doubt that it perished from the same causes as its father. I sat with her and the nurse by the fire very late into the night. Whilst I was there the children did not wake, nor seem in any way disturbed, and at length I thought I might myself seek some repose. When I returned about six o'clock in the morning, another child had severe cramps with some sickness, and while I stood by the bedside, it died. Then, later, the third and eldest child had all the characteristic symptoms of cholera and perished in one or two hours. In the course of the day the mother likewise suffered from a severe and rapid succession of the characteristic symptoms and died, so that within twenty-four hours the whole family was extinct, and it was not known that any other case of cholera had occurred in Manchester or its vicinity.

*Source*: Carey (1987).

## Conclusion

Regimes lost legitimacy if the populations under their control were hungry too often. It was politically rational to provision populations in eighteenth- and nineteenth-century Europe; nation-building required it, as did the expansion of the franchise. Labour was required for industrialisation and warfare, and certainly because infectious disease could strike all social classes, investment in its elimination was rational. Interest in preserving life also reflected an increase in humanitarianism associated with the Enlightenment. After the French Revolution of 1789, all political leaders feared food riots and serious unrest associated with food price increases, because these could, and did, turn into revolutionary movements (see Chapter 3 for discussion of contemporary food riots). It is this political reality which explains state intervention to relieve famine, high food prices and disease; in the absence of that state commitment, hunger occurs. This brief review of the elimination of hunger in Europe illustrates that political will is a necessary factor in eliminating hunger and disease. It also establishes that the concept of entitlement, used to understand contemporary hunger, is applicable historically too.

## Key ideas

1 Hunger continues to exist because the political will to eliminate it is absent. Changes in political realities in Western Europe in the eighteenth and nineteenth centuries help explain why widespread hunger was eliminated there by the end of the nineteenth century.

2 Decisions made at all levels, from the international to the familial, help explain why some people eat well and others are hungry.

3 Although, in some cases, natural factors may trigger famine, hunger is a social phenomenon which can, therefore, be eliminated.

4 It is useful to analyse hunger by examining people's entitlements. Entitlements are the variety of means whereby any individual or group of people can command food; entitlements may be based on inherited capital, earned income, state provision, family or kinship provision, community obligations, or charity.

5 Frequently, individuals or groups have several types of entitlement (seasonal income and a little land, for example). Within all societies there are marked contrasts in entitlements.

6 Social change initiates changes in entitlements; sometimes change enhances people's entitlements and sometimes it reduces their entitlements. Frequently, social change improves entitlements for some at the expense of others' entitlements.

# Sierra Leone

## SWEET POTATO LEAF STEW                                    Serves 4

The Mende people in southeast Sierra Leone eat rice as their staple food. When it is prepared they spread it on a large round platter. A soup or stew, like this one, is served over the rice and then the people gather round to share the food from the common dish.

This is a very basic recipe and you can alter it by using beef or chicken instead of fish; or by substituting 1 cup/175g cooked beans (kidney beans, garbanzos/chickpeas or lentils) for the peanut paste.

Ruth van Mossel, Freetown, Sierra Leone

## INGREDIENTS
4 cups/500g green vegetable leaves (sweet potato or spinach), washed and chopped very finely
½ lb/225g smoked dried fish or 1 lb/450g fresh fish
½ cup/110g groundnut paste/peanut butter
3 cups/700ml water
1½ cups/360ml red palm oil*
1 large onion, chopped
1–2 fresh chillis, crushed, or 1–2 teaspoons chilli powder
salt and pepper

* Red palm oil can be found in Indian and Caribbean shops. If you cannot find it use peanut oil instead.

## METHOD
1 First soak the dried fish for 15 minutes and then take the bones out. Wash the fishmeat and place it in a medium-sized saucepan. Pour the red palm oil over and add 1 teaspoon of salt. Heat up gently and then let simmer for 5 minutes.
2 Now add 1 cup/240ml water to the simmering mixture as this cooks and softens the dried fish. Put the lid on.
3 Next put in the onion, chillis or chilli powder, some more salt and 1 cup of water. Continue to simmer until the fish is tender, about 15 minutes.
4 When it is cooked, remove the lid from the pot and lay the chopped green leaves on top of the fish mixture. Drizzle a little palm oil on before replacing the lid and then leave it to simmer again.
5 In a bowl, stir about 1 cup/240ml water into the peanut butter to make a thin, smooth paste, adding more water if necessary, and pour this into the stew. Stir well, adjust the seasoning, and put in more water if the stew is too thick.
6 Serve over cooked rice and accompany the dish with an array of fresh fruits.

# The contemporary nature and extent of hunger

## Terminology

### Chronic and acute hunger

Before launching into further debates about the causes of world hunger, two preliminary exercises are essential. The first is to appreciate the difference between chronic and acute hunger and the second is to understand the nature and extent of both. Most people are vague about the distinctions between acute and chronic hunger and indeed the boundary is very problematic; however, it is important to establish the difference. Debates about world hunger are concerned about both acute and chronic hunger but the world's media grant the problem of acute hunger, that is, famine, more attention. Ask students to mark on a map the famine-prone regions of the world and most will identify Ethiopia, Sudan, Bangladesh and perhaps Somalia. Asked to identify where most of the world's hungry live they fare less well. Why?

Famine makes a better story and the images are more lasting and moving; it is a very visible tragedy. The day-to-day debilitations associated with chronic hunger are less amenable to headlines but have been described as 'the insidious sabotage wrought' on millions of children in the developing world. Famines unfortunately still claim many lives, but the absolute numbers affected are less than those who are chronically hungry, who suffer debilitation from malnourishment. Uvin (1994) estimates that 13 million people die each year from extreme malnutrition and hunger-related causes, some 35,000 per day; three-quarters of them are children. Ten per cent of these deaths are from famine and

90 per cent are caused by chronic, persistent hunger. Chronic malnutrition breaks down resistance to even the mildest diseases like the common cold and diarrhoea.

## Famine

> Famine is like insanity, hard to define, but glaring enough when recognised.
>
> (quoted in Devereux, 1993)

> Famine is one of the most powerful, pervasive, and arguably one of the most emotive, words in our historical vocabulary, and that in itself makes it all the more difficult to isolate its meaning and wider significance.
>
> (Arnold, 1988, 5)

I have already used the term 'famine' in Chapter 1 and would be surprised if any readers asked themselves 'What exactly does she mean when she writes of famine?' Trying to define exactly what constitutes a famine is more complicated than would first appear. Arriving at a universally accepted definition is impossible, but because it is a concept around which institutional responses are geared and because it is, at some stage, distinct from generalised situations of chronic hunger, it is essential to review attempts at its definition. This task is also useful because it helps indicate the difference a word makes and how complex apparently simple concepts may be upon investigation.

A famine is an exceptional event, in different societies at different times; they may be very rare or more frequent, but they are never the normal state of affairs. Famines certainly happen and victims and observers recognise them as specific 'events', but commonly their beginnings and endings are problematic because they rarely happen without warning and their consequences continue when more normal circumstances are re-established. One way of identifying famine is by the associated increases in mortality. Sometimes, however, mortality increases are not as drastic as might be expected, because populations often have famine prevention strategies and domestic or international relief efforts may avert a massive mortality crisis.

Famines are always associated with horror and confusion. They have been interpreted as signs from God, as divine interventions in response to human wickedness, and they sometimes precipitate a change in government. In famine circumstances, conventions are disregarded and systems of law and order collapse; they are correlated with awful social dislocation, migration, family break-ups, the breaking of taboos. Everything considered 'normal' is threatened. Every famine, then, has its own character and an exact definition is impossible but the following is a useful working definition:

famines may be distinguished from the background of grinding poverty because in famine years there is a generalized crisis: in peasant societies large numbers of people may migrate to the cities, markets may be disrupted, labourers' wages may fall rapidly, the price of the staple food may rocket and organisations who care for and/or bury the destitute find that their task is suddenly unmanageable.

(Open University Social Studies Course D208, Unit 24, p. 4)

## Malnutrition

Perhaps the concept 'malnutrition' will prove more readily captured by a simple measurement of calories consumed per day. In fact this is not the case and exact definitions of what constitutes 'malnourished' or 'undernourished' are also fraught with problems. Again, while we must accept that an absolute definition is bound to elude us, it is crucial that an attempt be made and a standard definition employed. That such definitions vary is illustrated by the range of estimates available from international and national agencies faced with measuring malnourishment globally or nationally. A recent estimate (1992 by the FAO/WHO) put the number of malnourished in the world at 786 million. These malnourished people may suffer from any one of the following types of malnutrition: dietary deficiency; secondary malnutrition; undernutrition. These distinctions are important and are considered below.

Micronutrient deficiencies each singly constitute a brake on socio-economic development and mostly are combined in synergistic action to the detriment of the world's already underprivileged groups.

(Uvin, 1994, 21)

Deficiencies in any of the micronutrients necessary for good health, e.g., iron, iodine, vitamin A, the other vitamins, major and minor trace elements, are manifest in various ways and are known as dietary deficiency malnutrition. Because this form of malnourishment is insidious it is sometimes known as 'hidden hunger'. Although it is less visible than protein–energy undernutrition (see p. 25), if a diet lacks one or more essential nutrients then the consequences for health can be very serious. Since the early 1990s, considerable effort has been devoted by the major international donors to addressing this problem. Solving the problem of dietary deficiencies is economically cheap and politically uncontentious. Many people still suffer from dietary-deficiency and related disease, however. Some of the most serious such diseases are described below with an indication of the numbers of children affected, based on UNICEF estimates; remedies and progress are detailed (UNICEF, 1994).

Iodine deficiency is a very serious form of dietary deficiency. Worldwide, a total of 26 million people are estimated to be brain damaged by lack of iodine in the diet and as many as 600 million are physically or mentally affected in some way by the absence of iodine in their diets. Iodine-deficiency disorders are the world's biggest cause of preventable mental retardation. Unlike many other forms of malnutrition iodine deficiency can be solved very simply by adding iodine to common salt. Because salt is used in cooking and to preserve foods, sufficient amounts are absorbed by the general population. The World Health Organization (WHO) has reasonable confidence that by 2000 targets set at the World Summit for Children in 1990 will be met. That target was that all countries affected by iodine deficiency-related problems would have reached 95 per cent salt iodisation.

A second common and serious deficiency is of vitamin A. Lack of vitamin A causes approximately 500,000 children to lose their eyesight every year and half of these children do not survive. Approximately 230 million children have lowered resistance to disease because of milder forms of vitamin A deficiency; and the consequence is death rates that are commonly 20 to 30 per cent higher than in children whose vitamin A intake is adequate. Common causes of death associated with this are diarrhoea and measles. Improving vitamin A in the diet will reduce, not eliminate, the toll from these two diseases.

Again the remedy is not complicated or politically contentious. A small intake of leafy green vegetables every day, or a tablet which costs 1p three times a year, or vitamin A added to sugar or cooking oil would eliminate this problem. The World Summit for Children in 1990 argued that 'improving vitamin A intake was one of several obvious, powerful, low-cost strategies with the potential to reduce illness, blindness, and death among the children of the developing world' (UNICEF, 1995, 6). By the end of 1995, it is estimated that two-thirds of all children at risk will have a diet adequate in vitamin A.

Iron deficiency leads to iron anaemia and is particularly common in women of menstruating age (Box 2.1). Even small amounts of iron deficiency are associated with impaired work performance, damaged learning ability and dysphagia. Anaemia increases susceptibility to illness, pregnancy complications and maternal death. Modifying or supplementing diets only slightly could reduce this form of deficiency markedly.

malnourishment is not always a simple consequence of inadequate food supply.

(Grigg, 1993, 11)

Box 2.1

## Women and anaemia in the developing world

Between 20–45% of women of child-bearing age in the developing world do not eat the WHO recommended 2,250 calories a day under normal circumstances, let alone the extra 285 a day they need when they are pregnant.

(Smyke, 1991, 17)

Anaemia is a particularly serious problem for women in the developing world because women need more iron than men to lead healthy productive lives. Anaemia may be caused by a lack of iron in the diet but it is often exacerbated by the presence of hookworm and diseases like malaria. An anaemic women is tired and cannot be as productive, and an anaemic woman is more prone to disease; both have serious implications for herself, her children and the community. Anaemia is a particularly good example of the interactions between diet and disease. Unless her diet is adequate in iron, each pregnancy depletes her supply of iron, which means that she is more vulnerable to infections, disease and death and less able to deal with the demands of breast-feeding and the next pregnancy. An anaemic woman is likely to have low-birthweight babies who are liable to become undernourished children prone to infection. Malaria has a dangerous association with anaemia. Its incidence is higher among young pregnant women. Both can be identified early and treated, so it is vital that young women are screened for both.

A variety of social factors have negative implications for women's nutritional status and health: poverty; low social status; discrimination against girls; none/limited family planning services; restricted access to education; poor primary health provision; early pregnancy. At different stages of her life cycle a poor woman in many developing countries is at risk from different nutritionally related health problems (see Figure 2.1). Infant girls born to poorly nourished mothers are low in birthweight because of growth retardation in the womb; at greater risk from diarrhoea; more likely to suffer from learning difficulties. As a teenager and young woman this child will continue to have learning difficulties and stunted growth and will have a narrow pelvis and may have delayed menarche. As a potential mother she is at risk from vitamin and mineral deficiencies and their attendant health problems and especially now she will be liable to anaemia. As a poorly nourished young mother she may have anaemia, less breast milk and more birth complications, and she risks bone deformity.

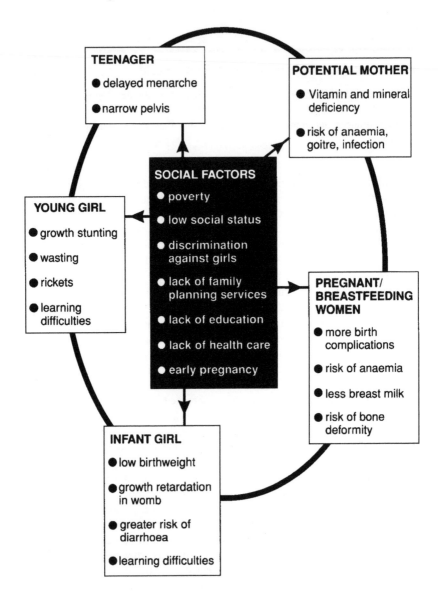

**Figure 2.1** The vicious circle of female malnutrition
*Source*: Smyke (1991, 18)

Another type of malnutrition is known as secondary malnutrition and is associated with the complex interactions between diet and disease or illness. Some diseases are particularly virulent in undernourished populations. The main killer of undernourished children is diarrhoea, but other major killers associated with undernourishment are pneumonia, influenza, bronchitis, whooping cough and measles. Some conditions mean that the body cannot adequately exploit the food that is available and this results in secondary malnutrition. The most common causes of secondary malnutrition are diarrhoea, respiratory illnesses, measles and intestinal parasites. These conditions are associated with some, occasionally all, of the following symptoms: loss of appetite; poor nutrient absorption; diversion of nutrients to parasites. Deaths due directly to undernutrition are limited, the bulk of deaths being from diseases associated with undernutrition. Undernourished people are more vulnerable to diseases.

Diarrhoea and pneumonia remain two of the most serious causes of death among children in the developing world; each of these claims approximately 3 million young lives per year, or nearly 50 per cent of all deaths under the age of five. They are both cause and consequence of malnutrition and illustrate the complexities of the interactions between unhealthy living conditions and malnutrition. Sometimes they also reflect the marketing success of major multinationals (Box 2.2).

Box 2.2

---

### The breastfeeding debate

Imagine a new 'dream product' to feed and immunise everyone born on Earth. This product requires no storage facilities or delivery and helps mothers plan their families and reduces their risk of cancer. What is this product? Human breast milk, available to all at birth, and yet breast-feeding is declining in both rich and poor countries. Despite over-whelming scientific evidence that human breast milk is superior to any of the infant formulas, even if they are mixed with clean water and given from sterilised bottles, formulas continue to usurp breast milk as the primary food in the first months of the newborn's life. Technological advances in packaged foods during the 1950s made it possible to offer breast milk substitutes to women in the industrialised world. Aggressive marketing, free samples, and intensive promotion through hospitals and

Box 2.2 (*continued*)

health centres help explain why infant formulas rapidly usurped breast-feeding in the West. Potential markets in the developing world were targeted and exploited next. As the use of formulas was seen to be 'modern' and promoted by health professionals, its success in the developing world was assured.

Every year over 1 million infants die, and millions more are impaired, because they are not adequately breastfed.

(UNICEF, 1994)

As poor households in the developing world are not equipped to sterilise bottles and do not have access to safe water to dilute the formulas, babies are exposed to a greater likelihood of malnutrition and diarrhoeal diseases.

A global campaign by health organisations and citizens' groups led to adoption of the International Code of Marketing of Breastmilk Substitutes by the World Health Assembly in 1981. However, the 1980s saw bottle-feeding continue to increase and breastfeeding continue to decline. More recent declarations from the United Nations and governments have stressed the unparalleled benefits of breastfeeding; the Convention on the Rights of the Child, entered into in 1990, made it a legal obligation of states to provide mothers and families with the knowledge and support needed for breastfeeding; the Innocenti Declaration, signed by thirty-two governments and ten United Nations agencies in 1990, recognised the need for global support for breastfeeding; a consortium of major international non-governmental organisations (NGOs) formed the World Alliance for Breastfeeding Action (WABA) in February 1991; in 1991 the International Association of Infant Food Manufacturers promised to stop supplying free and low-cost breast milk substitutes to hospitals and maternity centres throughout the developing world by the end of 1992.

*Source*: UNICEF (no date).

Where people are living without access to clean water and adequate sanitation, illnesses gradually erode their health and ability to exploit the food they do eat. To eradicate malnutrition, then, people must be granted access to both clean water and safe sanitation. The International Drinking Water Supply and Sanitation Decade (1981–90, organised by the UN) saw the proportion of families with access to safe drinking water rise from 38 to 66 per cent in Southeast Asia, but although improvements were achieved in Africa, access remains at a low of 42 per cent. India achieved considerable success with an increase from 30 to 80 per cent of rural people having access to clean water. The statistics are encouraging but are not without problems of interpretation.

The final form of malnutrition is the one that most people think of when you mention malnutrition; it is undernutrition. Undernutrition occurs when an individual's diet is short of calories and/or protein necessary for normal growth, body maintenance, and the energy required for normal activity. This type of hunger is most common among the poorest populations in the developing world. There are a number of symptoms of undernutrition. Undernourished populations are more vulnerable to infection and disease than those that enjoy a diet that is adequate in calories, proteins and nutrients. Several physiological symptoms suggest that an individual is, or has been, undernourished. These are expressed as a comparison to a reference population.

Two clear indications of undernutrition are low birthweights and high infant mortality rates (IMRs). IMR is the number of deaths of infants under one year old per thousand live births. Compare the IMRs of the UK (7.1), the USA (8.6) and Japan (4.4) with those of India (91) and Bangladesh (116). Remember that these figures are means (averages) and conceal a great deal of domestic divergence from the mean. For example, IMRs in poor inner city areas of the USA are comparable with those found in Bangladesh. Low birthweight rates for African-Americans place them in seventy-seventh position worldwide, after Ivory Coast, Senegal and Lesotho. Even the average US birthweight places it behind some countries in the developing world. When the IMR falls below 50 persistent hunger is no longer considered a major issue (Uvin, 1994).

Three other indications of undernutrition are often used by researchers in surveys to establish its extent. Low height for age may be associated with undernutrition in the past even if diet is adequate at the present. Tables are computed to establish whether an individual is significantly small for their age. Calculations allow for normal height variations. Then if an individual has a low weight for their height, this probably indicates that they are currently undernourished. A low weight for age may indicate past or present under-nutrition. A useful guide to nutrition in young females is their age when they get their first period. A delayed age of menarche (first menstruation) indicates low levels of calorie intake. The female sex hormone, oestrogen, is produced from cholesterol, a fat, so reduced calorie intakes are associated with both late menarche and infrequent periods.

The most extreme forms are exemplified by nutritional conditions known as marasmus and kwashiorkor, which are fatal without quick, intensive medical intervention. The word 'kwashiorkor' comes from Ghana and actually means 'the evil spirit that infects the first child when the second is born'. It is clear how this name arose. When a mother becomes pregnant with a second child she weans her first child, and its diet of protein-rich mother's milk is replaced by a protein-deficient diet. Children suffering from kwashiorkor have swollen bellies from oedema and some or all of the following symptoms of under-nourishment: stunted growth; loss of hair colour; patchy and scaly skin; ulcers and open sores. They sicken easily and are weak, fretful and apathetic.

When the body has insufficient proteins available to meet all its needs, it reduces body maintenance and prioritises the most vital functions. So the hair and skin are neglected in favour of maintaining the heart, lungs and brain tissue. Many of the antibodies are also degraded in the body's attempt to build the vital organs. This precipitates a downward spiral because as antibodies are reduced the body becomes vulnerable to infection and readily contracts dysentery, a disease of the digestive tract. Dysentery causes diarrhoea, leading to the rapid loss of nutrients, which worsens the protein deficiency, which increases the probability of a second or third attack of dysentery.

Marasmus, a wasting disease, is another extreme condition associated with severe undernutrition. When the body does not get enough calories (energy) it breaks down protein to use as energy, so many children with marasmus get protein-deficiency problems too. Marasmus may occur in adults or children but if it occurs within the first two years of life, brain development is impaired. The symptoms of marasmus are shocking because children look aged and lack all the normal interest and energy of infants. Children are usually sick because their resistance to disease and infection is low and their muscles are wasted; this includes their hearts, which are muscles too. Their metabolism is slow and they have very little fat to keep them warm. Marasmus is most likely to occur in populations suffering extreme poverty, whose access to calories is very inadequate.

## The contemporary extent of hunger

The opening section of this chapter describes some of the difficulties which surround the terminology of hunger. This section reviews the type and quality of the statistics employed to map hunger before describing its distribution in the 1990s. The political context of accounting needs a mention first. Different interest groups at the international, national or regional level may be best served by manipulating statistics to exaggerate or diminish the incidence of hunger. To classify a situation as 'famine' may result in the provision of international relief, which may be welcomed by national governments, or the use of the term may be eschewed because of the implications it has for national policy (Jowett, 1987 for the Chinese case in 1958). Similarly governments, careful of their reputations domestically and in the international arena, often deny the seriousness of malnutrition; frequently the well-being of specific minority groups is politically charged.

An essential first step is to review the statistics available to analysts. How may we determine the incidence of hunger? The Food and Agriculture Organization (FAO) is the main source for data on world hunger. It begins by calculating a country's food supply by totalling all food produced and imported into a country, and subtracts carryover stocks, loss (due to pests, etc.), seeds and exports. These totals are then converted into grain equivalents, adjusted to reflect the distribution of food between households and compared with the energy needs of the population. The FAO uses a Basic Metabolic Rate (BMR – the caloric expenditure of an immobile body in a warm environment) of 1.54, which allows for light activity but not manual labour. The resulting estimates of malnourished persons worldwide between 1970 and 1990 suggest that the absolute number of malnourished has declined since 1975, from 976 to 786 million. So, despite an additional 1.1 billion people in the world, the incidence of hunger in the world has declined. However, that is no comfort to those 786 million still acutely or chronically hungry.

These statistics must be disaggregated to map the geography of world hunger. Where do these 786 million people live? Table 2.1 shows the number of poor and hungry in the various world regions. The greatest absolute number of hungry people, an estimated 277 million in 1990, live in South Asia; most of these people live in India, Bangladesh and Pakistan. In China an estimated 189 million are chronically underfed; East Asia has 74 million, while South America has 38 million, Middle America 20 million and the Near East and North Africa 12 million. These absolute numbers confound the popular image of sub-Saharan Africa as being the worst region for hunger. Acute hunger there may be higher and subject to drastic fluctuations but South Asia is home to

**Table 2.1** The percentage of chronically underfed, by region

| Year | SSA | NE/NA | MAm | SAm | SA | EA | China | All |
|------|-----|-------|-----|-----|-----|-----|-------|-----|
| 1970 | 35  | 23    | 24  | 17  | 34  | 35  | 46    | 36  |
| 1975 | 37  | 17    | 20  | 15  | 34  | 32  | 40    | 33  |
| 1980 | 36  | 10    | 15  | 12  | 30  | 22  | 22    | 26  |
| 1990 | 37  | 5     | 14  | 13  | 24  | 17  | 16    | 20  |

Source: Uvin (1994)

Notes: Sub-Saharan Africa (SSA); Near East and North Africa (NE/NA); Middle America (MAm); South America (SAm); South Asia (SA); East Asia, excluding China (EA).

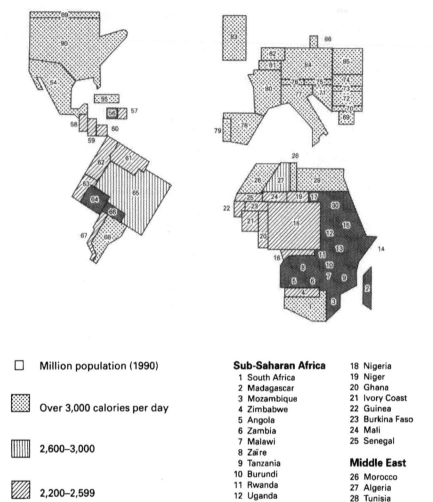

| | Million population (1990) | **Sub-Saharan Africa** | 18 Nigeria |
|---|---|---|---|
| | | 1 South Africa | 19 Niger |
| | | 2 Madagascar | 20 Ghana |
| | Over 3,000 calories per day | 3 Mozambique | 21 Ivory Coast |
| | | 4 Zimbabwe | 22 Guinea |
| | | 5 Angola | 23 Burkina Faso |
| | | 6 Zambia | 24 Mali |
| | 2,600–3,000 | 7 Malawi | 25 Senegal |
| | | 8 Zaïre | |
| | | 9 Tanzania | **Middle East** |
| | | 10 Burundi | 26 Morocco |
| | 2,200–2,599 | 11 Rwanda | 27 Algeria |
| | | 12 Uganda | 28 Tunisia |
| | | 13 Kenya | 29 Egypt |
| | Less than 2,200 | 14 Somalia | 30 Sudan |
| | | 15 Ethiopia | 31 Turkey |
| | | 16 Cameroon | 32 Syria |
| | | 17 Chad | 33 Saudi Arabia |

**Figure 2.2** Distribution of sample countries by per capita daily calorie intake

*Source*: Dyson (1996, 40–1)

*Note*: Areas of countries on the map are proportioned to the size of population in 1990

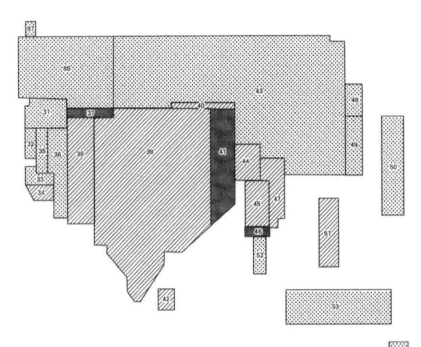

| | | | |
|---|---|---|---|
| 34 Yemen | 48 North Korea | 64 Peru | |
| 35 Iraq | 49 South Korea | 65 Brazil | |
| | 50 Japan | 66 Bolivia | |
| **South Asia** | 51 Philippines | 67 Chile | |
| 36 Iran | 52 Malaysia | 68 Argentina | 80 France |
| 37 Afghanistan | 53 Indonesia | | 81 Belgium |
| 38 Pakistan | | **Europe/FSU** | 82 Netherlands |
| 39 India | **Latin America** | 69 Greece | 83 UK |
| 40 Nepal | 54 Mexico | 70 Bulgaria | 84 Germany |
| 41 Bangladesh | 55 Cuba | 71 Yugoslavia (former) | 85 Poland |
| 42 Sri Lanka | 56 Haiti | 72 Romania | 86 Denmark |
| | 57 Dominican Rep. | 73 Hungary | 87 Sweden |
| **Far East** | 58 Guatemala | 74 Czechoslovakia (former) | 88 Former Soviet Union |
| 43 China | 59 El Salvador | 75 Austria | |
| 44 Myanmar (Burma) | 60 Honduras | 76 Switzerland | **N. America/Oceania** |
| 45 Thailand | 61 Venezuela | 77 Italy | 89 Canada |
| 46 Cambodia | 62 Colombia | 78 Spain | 90 USA |
| 47 Vietnam | 63 Ecuador | 79 Portugal | 91 Australia |

**Table 2.2** The absolute number of chronically underfed, in millions, by region

| Year | SSA | NE/NA | MAm | SAm | SA | EA | China | All |
|------|-----|-------|-----|-----|-----|-----|-------|-----|
| 1970 | 94 | 32 | 21 | 32 | 255 | 101 | 406 | 942 |
| 1975 | 112 | 26 | 21 | 32 | 289 | 101 | 395 | 976 |
| 1980 | 128 | 15 | 18 | 29 | 285 | 78 | 290 | 846 |
| 1990 | 175 | 12 | 20 | 38 | 277 | 74 | 189 | 786 |

*Source*: Uvin (1994)
*Notes*: Sub-Saharan Africa (SSA); Near East and North Africa (NE/NA); Middle America (MAm); South America (SAm); South Asia (SA): East Asia, excluding China (EA).

most chronically malnourished people. It is important to stress that the map changes shape if a different statistic is employed, as shown in Table 2.2. If we map the percentage of total populations estimated to be hungry, then indeed, sub-Saharan Africa emerges as the worst case, with 37 per cent of the population underfed, while in South Asia it is 24 per cent and in China an estimated 16 per cent.

Figure 2.2 is a useful overview of patterns of world hunger but must be used cautiously because it shows aggregate figures of food availability; it ignores discrepancies of distribution. The map confirms that hunger is most serious in South Asia and sub-Saharan Africa. Of course, within each region and country specific groups of people suffer a higher incidence of hunger than others. The purpose of this text is to understand why some groups suffer from hunger and others do not. Within any country there are sharp contrasts between those who enjoy a decent diet and those who do not. Are there any introductory generalisations which can be made to summarise the circumstances of those who are hungry? The answer is that there are some social groups who have a higher probability of suffering from poverty and hunger than others.

## Conclusion

Regional categorisations help us begin to understand the patterns of world hunger. At the global level, while hunger persists in the 'rich' world amidst some of the wealthiest populations, the great majority of the world's hungry live in the developing world. The statistics in Table 2.3 reflect the contrasts that exist within the South. At the sub-national level some regions are particularly associated with hunger: the Sahel region of West Africa and the northeast of Brazil (Box 2.3) are notorious examples. Understanding these regional contrasts is one of the main objectives of this analysis. A second category highly correlated with hunger is that of the refugees and displaced people, who are considered in more detail in Chapter 7.

**Table 2.3** GNP per capita and selected social indicators, selected countries, late 1992

| Country | GNP<sup>a</sup> | IMR<sup>b</sup> | Life exp. (yrs) | Daily calories<sup>c</sup> |
|---|---|---|---|---|
| China | 1,950 | 44 | 68.5 | 2,729 |
| Bangladesh | 1,230 | 108 | 55.6 | 2,019 |
| India | 1,230 | 82 | 60.4 | 2,395 |
| Indonesia | 2,950 | 58 | 62.7 | 2,755 |
| Egypt | 3,540 | 67 | 63.6 | 3,336 |
| Guatemala | 3,330 | 48 | 64.8 | 2,255 |
| Thailand | 5,950 | 37 | 69.0 | 2,443 |
| Turkey | 5,230 | 65 | 66.5 | 3,429 |
| Peru | 3,300 | 64 | 66.0 | 1,883 |
| Brazil | 5,240 | 58 | 66.3 | 2,824 |

Source: Based on data in UNDP (1995)

Notes: <sup>a</sup> Real GNP per capita in US$.
<sup>b</sup> Infant mortality rate (IMR): number of infants who die before reaching one year of age, per thousand live births in a given year.
<sup>c</sup> Daily calorie supply per capita: calorie equivalent of net food supplies divided by the population.

**Box 2.3**

### The northeast of Brazil

And so we say that infants are like birds – here one moment, flying off the next.

The past thirty years has seen a slow but marked improvement worldwide in the number of children surviving beyond their fifth birthday. In 1960, out of every 1,000 children born in Brazil, 181 died before the age of five; by 1992 the number had been reduced to 65. But the differences between countries are sharp; between regions in the same country they can be sharper still; in Brazil they are among the sharpest in the world. The following is a selected list of under-five mortality rates per 1,000 live births, 1992:

| | |
|---|---|
| Ireland (lowest in the world) | 6 |
| USA | 10 |
| Cuba | 11 |
| Argentina | 24 |
| Brazil | 65 |
| Northeast Brazil, 1987 | 116 |
| India | 124 |
| Niger (highest in the world) | 320 |

Box 2.3 (*continued*)

The northeast of Brazil is notorious for its poverty. This has its roots in the history of the region and the way it was incorporated into the world economy as a Portuguese colony in the seventeenth to the nineteenth century, but the persistence of poverty requires reference to contemporary international and national circumstances. The colonial economy established by the Portuguese from the seventeenth century onwards was based on the production of sugar. It was organised as a plantation system, where the land was held in extensive estates by a few landowners, of Portuguese descent, who lived in the *casa grande*. Labour was acquired by importing slaves from Africa. This area of Brazil was incorporated as a part of the infamous Triangular Trade, the Atlantic slave trade. This element of the emerging world economy was largely managed by British, French, Dutch, Spanish and Portuguese interests, which traded Africans and the products, raw and manufactured, of the tropical plantation economies – tobacco, sugar, coffee, cocoa, indigo. This trade lasted approximately 400 years, and some 10 million slaves, the largest forced migration in history, were transported from Africa to the Americas. An estimated 50 per cent of them went to Brazil, a great many to work in the sugar plantations of the northeast in the hinterland of Recife. The imported labourers lived some distance from the *casa grande* in the *sezala*, a series of connected one-room slave quarters. Each plantation had a mill to process the sugar cane before export to Europe.

Until the twentieth century, cultivation and processing processes were very inefficient, labour intensive and environmentally destructive as trees were ruthlessly felled to fuel the mills and the sugar cane fields were exploited until the soil was exhausted. A new type of system emerged at the beginning of the twentieth century, when modern sugar mills were built beside the railways. The old-style feudal plantation barons lost some of their dominance as they lost control of mill processing and became instead suppliers of cane. There were serious repercussions for the traditional peasant class of the region – tenants, sharecroppers and squatters – who had existed by selling their labour and cultivating small plots. In the 1950s a series of evictions removed these people from the estates, and they became landless labourers who moved to live in rapidly constructed squatter communities. These workers remain a readily available, and readily exploited, labour force for cane cultivation and processing. Labourers and peasants in the northeast were not passive in

Box 2.3 *(continued)*

the face of their ruthless exploitation and unrest was common. Peasant and worker unrest was serious in the 1960s but was crushed by the military junta then in power.

What this case study illustrates is the history of poverty in the northeast of Brazil and the history, therefore, of its high infant mortality rates and the generalised malnutrition which typifies the region. It also suggests that intra-national contrasts in nutrition and IMRs may be greater than international differences. Finally, it is a dramatic and early illustration of the links that integrate producers and consumers in the world food system. The Portuguese colonists developed a plantation economy using slave labour to produce sugar, which was, to the world economy of the seventeenth and eighteenth centuries, similar to the petroleum trade in the world economy of the 1960s. The market was insatiable and profits were worth killing and dying for. Initially the wealthy, and increasingly the poorer classes in Europe became addicted to sugar and until the development of sugar beet, a temperate crop, in the twentieth century, this meant dependence on tropical production. The wealth that its trade generated, however, was not distributed equitably. The land-owning system in Brazil ensured that a few enjoyed great wealth and the mass of workers lived in poverty and insecurity. They still do in the northeast of Brazil, as reflected in the awful infant mortality rates. This region has a fickle rainfall regime and the legacy of this system of production has exacerbated problems by leaving a miserable landscape – depleted soils, slopes denuded of trees and soil erosion.

*Source*: Some of this is based on the work of Nancy Scheper-Hughes and her book *Death without Weeping*, excerpts of which are available in the *New Internationalist*, no. 254, April 1994.

A third characteristic of hungry populations is related to gender. Women suffer a higher incidence of hunger than male populations in some regions.

Numerous studies reveal that where food is in short supply it is not distributed on an equitable basis within the family. The nutritionally vulnerable are most often young children (particularly females), pregnant mothers, and the nursing mother.

(Warnock, 1987, 10)

The influence of gender upon food entitlements is the theme of Chapter 5. Minority populations too suffer a greater incidence of hunger in many countries than the majority group. A fifth factor is age related. At either end of the life cycle the incidence of hunger is higher than for the general population – infants and old people suffer from hunger more than others. Why these categories of people are more vulnerable to hunger is the focus of the rest of this text.

## Key ideas

1 Although famines still claim too many lives, the majority of the world's deaths due to hunger are from chronic hunger and associated diseases.
2 There are approximately 800 million hungry people in the world; the majority of these people live in Asia.
3 There are several forms of malnutrition; undernutrition, micronutrient deficiencies and secondary malnutrition. Each of these types of malnutrition is addressed by different policy interventions. The easiest to eliminate is micronutrient deficiencies.
4 There are millions of people in the South who eat good, diverse and healthy diets; try some of the recipes given at the beginning of every chapter! There are stark contrasts in diets *within* countries as well as *between* them. There are hungry populations in the world's wealthiest countries and in all countries in the South some people eat very well.
5 Within countries, extreme contrasts in food intake may exist between regions, classes, genders, and ethnic and age groups.
6 Infant mortality rates (IMRs) are a useful indication of well-being, including nutrition. When IMRs fall below 50 per 1,000, persistent hunger is considered to have been eliminated.

# Mauritius

## BEEF ROUGAILLE

This is a typical Creole dish. Lamb, chicken, goat or fish can also be used, but whatever meat you use it should be left after seasoning for 15 minutes. *Rougaille* is thought to come from French 'roux' meaning sauce and 'ail' – garlic. We are an Anglo-Mauritian family and this is a firm favourite with us.

Stella Bruce, Asha and Shona Brewer, Ilkley, UK

## INGREDIENTS

1½ lbs/675g lean beef, cut against the grain into very thin slices
2 tablespoons oil
3 onions, sliced
½ inch/1cm fresh ginger root, peeled and chopped finely
1 teaspoon fresh thyme
3 cloves garlic, crushed
4–6 chillis, split lengthwise
1 tablespoon fresh parsley, chopped
1 lb/450g tomatoes,
6 scallions/spring onions, sliced
1 tablespoon fresh cilantro/coriander, chopped

## METHOD

1 Sprinkle the salt and pepper over the meat slices, turning them so that the seasoning reaches each part evenly. Leave for 15 minutes.
2 After this time, heat the oil in a heavy pan and cook the meat for 1 minute, stirring frequently.
3 Take out the meat and put in the onions, ginger, thyme, garlic, chillis and parsley and cook for 3 minutes. Next add the chopped tomatoes and cook for a further 2 minutes.
4 Now return the meat to the pan and simmer for 15 minutes, or longer if required, until it is tender.
5 When the meat is cooked, add the chopped scallions/spring onions, half the cilantro/coriander leaves and more seasoning if required and continue to cook for a further 2 minutes.
6 Before serving, garnish with the remaining cilantro/coriander and then serve with rice or other accompaniment such as potatoes.

# 3
# International perspectives on global hunger

## Introduction

> The global agricultural and food system contains a fundamental contra-
> diction. On the one hand, it produces cheap and abundant food for the west.
> On the other, it creates a situation of regional socio-economic dependence
> that touches large segments of the population of developing countries and
> some segments of the population of developed countries.
>
> (Bonanno *et al.*, 1994, 256)

Analyses of hunger, famine or malnourishment are often written without
reference to global processes which have structured and continue to structure
the geography of food production and distribution. So undernourishment in
Bangladesh, for example, is analysed with reference to internal factors alone:
land ownership patterns, floods, urbanisation, etc. These national perspectives
are important (see Chapter 4) but they must be understood within a larger
geographical and historical frame. There are several compelling reasons why
an international perspective is required for an understanding of world hunger.

Historically constructed political, economic and social structures continue
to control access to and command over food, as well as decisions relating to
food production and consumption. Within countries, entitlement packages are
often based upon social and economic relations established under European
colonisation: the result of changes wrought between the sixteenth and the
nineteenth century. In external relations, some countries have a restricted
entitlement package with which to command food for their populations. At the

international scale, some countries are much more powerful politically and economically and can command food for their populations more effectively than countries in the South. As outlined in Chapter 1, the most critical requirement for any governing élite is to secure adequate food at prices which are affordable; if it fails to do this it loses legitimacy and risks being ousted. This rule holds for Chinese emperors in the past, new regimes in Eastern Europe in the 1990s, and contemporary governments in Africa and Latin America (see politics of structural adjustment, p. 43). This is not to argue that all relatively poor countries *will* contain hungry populations; not at all, it is simply to emphasise that an initial and vital constraint upon any country's ability to command food is its status in the international political arena, because this is where crucial policies which influence developing-world revenues are negotiated.

There is ample evidence that the world food system is becoming increasingly integrated, so that who eats and who does not, and what they eat or do not eat, is now influenced by global processes which are quantitatively and qualitatively different from global processes in the past. This chapter overviews inherited international processes that govern access to food before detailing some of the most recent transformations of the global production, distribution and marketing of foods, which is creating new winners and losers in the competition over the production and consumption of food.

## The changing geography of global food production and consumption, from the sixteenth to the twentieth century

The geography and diffusion of plants and animals in the ancient world, from Mesopotamia and other agricultural hearths, is a fascinating study, and trade in luxury foods existed in the centuries before European expansion: the trans-Saharan salt trade; the wine trade, which connected the Mediterranean regions to Northern Europe; the trade in olive oil; the trade in eastern spices to Europe. However, these transactions never represented a significant element of the diets of the majority of people, or even of the diets of the élite. The great bulk of food production and consumption was localised until capitalism emerged and diffused with the European conquests. The focus of this section is on some of the more important connections made between regions and peoples and their diets since European expansion and colonisation, that is, some important foundations of the contemporary global food system.

The discovery of the Caribbean as a sugar-producing system initiated a process of *producing* a food cheaply in one location and transporting it in large volumes to be *consumed* in another. The case study of northeast Brazil (see Box 2.3) is an early example of how an 'improvement' in the diets of some, the Europeans, had negative impacts for those of others. Although consequences were mixed in those regions devoted to the production of one or two crops destined for distant markets, the few generally prospered but not the many. Dietary demands in Europe and later other affluent markets in North America and Japan for luxury and 'exotic' crops are satisfied by sourcing foods in distant regions. Numerous crops could be cited to illustrate these connections, grown in diverse countries in the South: bananas, pineapples, coffee, tea, cocoa, etc.

However, it is not only the appearance of exotic luxury crops that illustrates early global food connections; other crops were socially and economically more important – those destined to become staple foods for the industrial populations of Europe. The introduction of the potato, a species native to the Andean region, in the seventeenth century transformed the diets of the poor in Europe and also their social relations. As industrialisation and proletarianisation proceeded in Europe the quest for basic foodstuffs became crucial. Wheat, processed into bread, became the basic foodstuff of the industrial poor. The main production centres were in Eastern Europe and subsequently in the 'New World', the prairies of North America, the pampas of South America, and Australia (Box 3.1). European urban demand in the nineteenth century was reflected in the massive expansion of wheat production in the American Midwest and the pampas of Argentina. Advances in transport technology were vital to all these developments: the railway in the late nineteenth century and refrigeration and air transport more recently.

Box 3.1

---

**Wheat and the pampas**

Wheat is one of the basic food staples in the North and increasingly it is replacing more traditional staples in the South, especially in urban areas, where it is assumed to be more modern. Wheat evolved in the Near East, from where its cultivation spread to Europe. From the sixteenth century onwards it was distributed by Europeans to new regions and in many it became the dominant cereal crop: the United States, Canada, Argentina and Australia. These countries and the EU are the dominant wheat exporters; that is, they produce large surpluses, unlike other major producers, where domestic consumption absorbs the bulk of production, e.g. China and India. The dominance of a few countries in world wheat

Box 3.1 (*continued*)

trade is cause for concern because serious downturns in their crop yields would precipitate price rises and, potentially, a food crisis. Wheat has several natural characteristics which help to explain its prominence in international trade. It has a high nutritional value and it is relatively easy to store, transport and process. Equally significant, however, are the historical circumstances which accompanied its adoption in the 'New World', where it became the main crop. Its diffusion in the Americas was facilitated by the appropriation of large tracts of land which were distributed to 'settler farmers' – farmers who enjoyed government incentives. Its dominance was assured when its production was allied to the massive beef-fattening business and with the triumph of the 'beef cultures'. The story of its introduction and ascendancy in Argentina illustrates the social context of its introduction and dominance.

In Argentina, native Indian populations resisted European colonisation, but by the nineteenth century the landscape, settlement and economy of the pampas had been transformed by European settlers who farmed to supply external markets. By 1830, large quantities of dried beef were being exported to the West Indies to feed the workers on the plantations (devoted to sugar or bananas). Next railways and port facilities were built, with British capital, and these transformed agriculture as they radiated out from Buenos Aires across the pampas to integrate this fertile region into the world market. Agricultural produce – grain and beef – flowed into the port, where it was processed before export to Europe. It became a very important source of cheap, protein-rich food for the growing industrial populations of Britain. In turn, Argentina received manufactured goods and migrants from nineteenth-century Europe. The introduction of canning technologies was the next important development and finally refrigeration facilitated the massive expansion of beef and sheep exports. In the twentieth century, the cultivation of wheat has replaced pastoral farming or been integrated into mixed farming systems through beef fattening.

Wealth was created in Argentina but it was enjoyed by a small élite. This economy was based on very inequitable land-owning structures. The pampas were divided into huge *estancias* owned by a few families, and cheap labour was provided by a flow of poor Europeans, who came to be tenants or labourers. Until recently, the 100 wealthiest cattle baron

Box 3.1 (*continued*)

---

families owned more than 10 million acres. The 2,000 largest farms covered one-fifth of the total area of Argentina. Peasants and ranch hands are hired to cultivate the wheat fields and to tend cattle and sheep but new technologies are marginalising many of them and massive migration to the cities is the result. The percentage of population classified as rural in Argentina dropped from 26 per cent in 1960 to 13 per cent in 1992. This process of agrarian change and consequent rural exodus is replicated with variations in much of the South (see Bernstein, Crow and Johnson, 1992 for an overview).

---

Until the early decades of the twentieth century, long-distance trade in food was restricted to bulky durable foods. After the Second World War, with the refinement of refrigeration techniques and canning processes, new options appeared and new geographies were generated. The combination of economic and social changes in the industrial countries – the creation of a wealthier class who demanded a diverse and protein-rich diet and technological developments in processing and transport – meant that new regions were integrated into the food system. Increasingly this new dietary pattern depends upon integrated networks of agri-food chains that deliver fresh fruits and vegetables from all over the world to the economically privileged. It is again important to draw attention to the fact that these general statements about dominant processes tell a partial story. Not everyone in the West eats well, and some people in the South eat an enviable diet. The character and implications of some of the most recent changes in agri-business and food chains are considered in more detail at the end of this chapter.

## International structures and national entitlements

The changing geography of food described above was predicated upon national and international structures which governed the movement of capital, trading relations, migration and labour processes, pricing policies, etc. These structures reflect unequal power relations within capitalism in the past and present. Rules were designed which prioritised the interests of the First World nations, the contours of the world economy emerged and patterns, once established, influenced future decisions and options. Patterns of trade, processes of production and advances in technology were dynamic but they were not chaotic. The structures which shaped the geography outlined above are reviewed next.

Since the emergence in Europe of a nascent capitalist world economy in the sixteenth century, a hierarchy of power relations has emerged which structures the production and distribution of goods, including food. This hierarchy is not static. The dominance of some states has declined and the power of others increased, but since the middle of the nineteenth century the states of Northwest Europe, North America and Japan have been relatively privileged *vis-à-vis* the countries in the South. The negotiating power of these states in institutions which govern trade helps explain why within their territories the threat of hunger is limited; hunger does exist in these countries, and in recession increases among the poorest sections of society, but as a national problem it is quantifiably different from the problem of undernutrition faced by most countries of the South. National statistics of calories per capita and infant mortality rates (IMRs), although they must be interpreted cautiously, do reflect real material contrasts between these national populations and their access to an adequate diet. Several examples are now considered to illustrate the relevance of an historically informed international perspective to analyses of world hunger.

Formal and informal colonial and neo-colonial relations were diverse and dynamic, but for our discussion one essential characteristic is relevant. Relations which evolved between the more powerful Northern political units and the Southern countries were fundamentally asymmetrical. As regions were integrated into a capitalist world economy between the sixteenth and the twentieth century, it was in terms initially and decisively disadvantageous to them (Knox and Agnew, 1994). Colonial and neo-colonial economies developed in response to European and later US needs. In most cases their role was to produce raw materials, including major food crops, and to buy manufactured goods produced in the industrial countries in Western Europe or the USA. By the end of the nineteenth century, whole regions of the world had evolved as export economies linked to the core regions of the industrial world. Places as diverse as Malaya (rubber), India (cotton), Egypt (cotton), Argentina (beef and wheat), Sri Lanka (tea) and Ghana (cocoa) became integrated as specialist producers for distant, more industrialised, and affluent markets.

The nature of integration into the capitalist world economy has meant that many are fatally dependent upon one or two export commodities. This makes their incomes vulnerable; a poor harvest or a fall in the world price of their exports can spell a financial crisis. The government will be strapped for revenues. Likewise, an increase in the price of basic food imports can effectively reduce the country's capacity to provide its citizens with food. Unfortunately, the terms of trade for agricultural commodity exports from the developing world declined by 40 per cent between 1980 and 1990, which meant that these countries' revenues have effectively declined while the price of imports increased. In the terminology of entitlements, some countries are dangerously exposed; slight fluctuations in the international arena can have serious repercussions for their entitlements and therefore their capacity to secure food in international markets.

## The debt crisis

> From 1970 to 1984, the total external indebtedness of developing countries rose from $64 billion to $686 billion.
>
> (Walton and Seddon, 1994, 14)

Since the mid-1970s, another factor has been eroding the ability of Southern nations to command food, that is the scale of debt repayments. A great deal of excellent analysis is available on the emergence and impact of the debt crisis. I will provide only a brief review to emphasise those elements of particular relevance to our understanding of hunger. As early as the mid-1960s, debt servicing was a problem for some developing countries – taking up 87 per cent of new lending to Latin America and 73 per cent of new lending to Africa. In other words, if the country borrowed £100, £87 went back immediately to repay debts incurred previously. These problems worsened in the 1970s and 1980s. What was the context of this crisis?

In the early 1970s, international financial institutions encouraged many governments in the Third World to borrow in the belief that borrowing to maintain economic growth through investment was the road to prosperity. Another factor which facilitated high levels of borrowing by Third World governments was that after the oil price rises of 1973 the coffers of the North's commercial banks were full and their managers eager to lend. Prospects for the borrowers changed for the worse after 1974, however, on several fronts: Third World countries were badly impacted by the oil price rise in 1973; interest rates on their borrowings increased; markets for Third World exports shrank as the world economy suffered a depression; and terms of trade for Third World exports deteriorated. A second oil price rise in 1979 exacerbated the problem on all fronts. Third World nations which did not enjoy oil reserves of their own were very dependent upon imported oil to fuel their industrialisation efforts. The intensification of commercial agriculture and the green revolution (Chapter 4) also demanded high inputs of petroleum-based products, such as fertilisers and pesticides. When oil prices increased, therefore, Third World borrowings also increased but at high interest rates and in conditions unfavourable to export sales. Repayments on accumulated debts become an acute financial crisis for the developing economies; Mexico's was such that it threatened to default on its loans in 1982. These circumstances had drastic implications for the entitlement packages of Third World countries. Simply, they were paying so much in loan repayments that their ability to purchase current commodities suffered. There were to be other serious ramifications of the debt crisis.

## The 'IMF food riots'

At the end of the Second World War, the Allies created the Bretton Woods institutions (BWIs) designed to govern capitalist international relations: the World Bank (WB), the International Monetary Fund (IMF) and the General Agreement on Tariffs and Trade (GATT). Understanding the role of these institutions is vital to analyses of contemporary patterns of global poverty and hunger. The IMF and the WB have had an interesting role since the 1980s. As the problems of debt increased, developing countries found it necessary to request loans from the IMF and the WB. By the 1980s, the IMF and the WB had concluded that structural adjustment policies (SAPs) should be implemented in the Third World. Before granting any further loans, therefore, or restructuring existing ones, the IMF and the WB insisted that SAPs be adopted. These policies varied but in all cases required that public (state) spending be reduced drastically and that exports be promoted. The theory was that by reducing spending and increasing earnings, the debts could be reduced more rapidly. A fascinating literature is available which examines the comparative political, economic and social consequences of these policies, and their differential impacts and potentialities, but there has emerged a consensus that, certainly in the short term, they have exacerbated poverty in the developing world. Too often in recent years it is the poorest segments of the population that have carried the heaviest burden of economic adjustment.

The programmes resulted in many of the poorest people suffering massive reductions in their entitlements overnight. A recent analysis asserts that the great majority of countries which have implemented SAPs have suffered losses in real income of 10 to 40 per cent during the 1980s – a drastic downward shift in entitlements for many people in the Third World. Sometimes a simple and direct relationship existed between adjustment policy and food prices; occasionally subsidies for basic foods were simply removed so that people suddenly had to spend a higher percentage of their income on food (Sudan,

Tunisia and Morocco). Adjustment policies often required currency devaluations, which effectively increased the cost of all imported goods, including food staples. Often adjustment policies fuelled inflation. Sometimes it was more indirect, with costs for health, fuel and education all increasing, but the end result was that whole classes of people saw their entitlements eroded. The cutbacks in Zimbabwe after the 1982 SAP had a markedly adverse effect upon child health, and surveys have confirmed a deterioration in nutritional intake, with an increase in child mortality between 1982 and 1984. Poor populations in the developing world were made to pay the cost of high interest rates consequent upon policies adopted in Western Europe and the USA and inappropriate bank lending in the 1970s. Implementation of adjustment has been uneven, partial and with immense human costs (Riley and Parfitt, 1994, 166). Not surprisingly, their implementation has been resisted.

> For developing countries around the world, the last two decades present an engaging historical problem: the reappearance of food riots and associated forms of popular unrest. Since the mid-1970s, an international wave of price riots, strikes, and political demonstrations has swept across the developing world in a pattern at once historically unprecedented and reminiscent of classical food riots best documented in European social history.
>
> (Walton and Seddon, 1994, 23)

Contemporary riots are different from those in European history because they are a product of the international political economy. While international elements may have been implicated in food riots in eighteenth-century Europe (diversion of state expenditures to military campaigns or generalised harvest failures, for example), the localised nature of food systems at that time meant that the sources of the shortages were regional not global. Riots in eighteenth-century Europe were against injustices *within* the nation-states. Contemporary riots are a violent response to the gross inequalities which characterise the international capitalist world economy. Resistance to the austerity programmes was manifest in the wave of riots which accompanied their implementation, known as 'IMF riots' or austerity protests. Many of these were linked specifically to the cost of food and can be accurately referred to as food riots. A typical example is from Madagascar after the externally imposed policy creating a free market in rice:

> Great hardship was in evidence in urban areas. There were reports in 1986 that poor families were selling their children in local markets as they were unable to feed them. During 1987 there was a famine in the south of

the country, rioting in the country's ports, attacks on Asian traders, and student strikes.

(quoted by Riley and Parfitt, 1994)

Resistance was obvious across Latin America too, some of the most serious being in Venezuela. In 1989, an extraordinarily violent series of riots swept across Caracas and sixteen other cities. A police inspector observed, 'It's a popular uprising. There are riots everywhere. They are furious.' Venezuela is a member of OPEC and exports oil but during the 1980s the price of oil fell and foreign exchange earnings dropped, so that by the end of the decade Venezuela had one of the continent's highest per capita debts. In 1981, 22.5 per cent of Venezuelans lived in acute poverty; six years later, that number was 54 per cent. Until 1989, the government had been complying with IMF conditions and implementing austerity measures but on the eve of the presidential campaign, the incumbent called the IMF 'a bomb that kills people with hunger' and pledged to limit repayments to 20 per cent of foreign exchange earnings. However, just three weeks after the president's inauguration, the government announced a structural adjustment plan which included elimination of price controls on basis foods and services; increases in the cost of gasoline and transportation; a public hiring freeze; a national sales tax and income tax reforms; interest rate increases; and currency devaluation. Riots ensued.

SAPs have had serious impacts and responses in Africa too. The case of Zambia is instructive. By 1986, the annual interest payments on the external debt represented 40 per cent of the government's budget compared with 15 per cent in 1980, and seven austerity programmes, various mixes of reduction in subsidies and price increases, were tried between 1975 and 1986. Popular responses were strong. One example of particular relevance happened in December 1986; violent riots occurred when the price of maize meal was doubled as a result of cuts in subsidies as part of an IMF austerity programme. Later, in July 1989, the price of maize doubled; it doubled again in June 1990, when a bag of maize meal (sufficient for a family for two weeks) cost $7.40, compared with the average weekly wage of a low-paid worker of $2. Interpretations within Zambia located the cause of Zambia's problems not in the external arena but in decades of corrupt domestic politics and nepotism. 'Thus a link between government incompetence, patronage, and drift – and economic hardship – was established and became politically salient, whilst the broader responsibility of the BWIs was minimised' (Riley and Parfitt, 1994, 150).

Simplistic distinctions between 'foreign' and 'domestic' interests are not easy to establish. Certainly many of the active participants – in the riots, strikes and demonstrations which have followed in the wake of SAPs – identify the agents of their suffering as international. Links are complex but Third World states are increasingly dependent upon the technical agencies and financial institutions of the international economy for loans, investment and, reluctantly, for policy choices that affect domestic food supply and entitlements. Food riots are attempts by poor and dispossessed people to command food and social justice in the face of these trends.

## The GATT

Another institution which is an important structure of the international capitalist world economy has been the GATT (modified to become the World Trade Organization, the WTO, in 1995). Although held to be an objective arbiter by its officials, many politicians and people in the developing world consider its policies to be biased in favour of the already advantaged industrialised affluent world. Its most recent deliberations are known as the Uruguay Round (UR) of the GATT.

GATT was designed to regulate world trade relations, establish quotas and tariffs, evaluate protectionist measures, etc. in the postwar era. Because of interests in the affluent world, agricultural sectors had been largely outside the GATT regulations until the UR. Since the surpluses of the 1970s and the 'farm wars' (between the EU and the USA) of the 1980s, however, circumstances have changed. Uruguay GATT debates stalled as the USA and the EU battled over agricultural policy, and Japan fought to protect domestic rice production against US imports. These conflicts were fuelled by attempts by the EU and the USA to maintain or increase their world market shares of agricultural trade, especially in wheat and meat. What have these conflicts over agricultural trading policy among the world's wealthiest nations got to do with explanations of hunger in less well-off regions?

The contemporary dominance of the USA and the EU in agricultural trade is based upon the dumping of highly subsidised food in Third World markets. What, you might ask, is wrong about sending cheap food from a 'surplus' area to a 'deficit' area? There may be several negative consequences. Cheap food imports undercut food production in the domestic agricultural sector and help maintain low incomes in this sector (Mies and Shiva, 1993). Rural producers suffer a loss of income and therefore their ability to command food. Imported foods are often associated with being 'modern' and 'Western'; marketing is very sophisticated and so imported foods become more popular than domestic-ally produced crops, despite often being less nutritious. Now GATT is liberalising trade in agricultural policies too, which has implications for producers and consumers everywhere, except in the most isolated regions.

Changes will have very mixed impacts in the developing world. Most experts expect the world price of cereals, milk products, sugar and beef to increase (because the EU must reduce subsidies on these goods, world prices will increase as supply falls). A recent analysis of the impact of the UR concluded that almost all individual sub-Saharan African countries lose. Trade reforms are just one symptom of the increased globalisation of agriculture and food systems more generally, considered next.

## Trends since the 1970s: globalisation and transnational corporations (TNCs)

> The internationalisation of agriculture and food is bound to make some of the most critical, positive as well as negative, contributions to the reconfiguration of global winners and losers.
>
> (Bonanno et al., 1994, 3)

I have established some of the vital international connections which have influenced diets since the sixteenth century. Since the 1970s, the globalisation of diets, and the agri-businesses upon which these diets depend, has intensified. Many argue that these changes signal a fundamental transformation, that a qualitatively new global food system has emerged. Alterations in social, political, economic and technological conditions are both cause and consequence of changes in global food systems; they will be crucial to the character of diets in the future (Box 3.2). Evidence of some of these momentous changes is considered next. I conclude this chapter with an evaluation of some of the potential consequences of these changes for people in different parts of the new food system.

> But it is in the spread of consumerism that multinationals have been most successful. In our age of instant communications corporate marketing messages blanket the globe. The dream of consumption without limits has seized the imagination of rich and poor alike. There is no corner of the earth so remote that it has escaped the notion that our identity is bound up in what we consume.
>
> ('Ellwood Multinationals and the subversion of sovereignty', *New Internationalist*, August 1993, p. 7)

Evidence of the intensification of globalisation (internationalisation) of the food system is everywhere. It is obvious in supermarket shelves in the affluent world and shops in the developing world. It is obvious in the hoardings which

Box 3.2

**Get fresh!**

Changes in social, political, economic and technological circumstances are both cause and consequence of the new global food regime. The most obvious symptom of this new regime is the dramatic increase in the long-distance trade in fresh products, as shown by the following data for fresh fruit and vegetable consumption in the United Kingdom, 1978–88, in £ millions.

| | |
|---|---|
| 1978 | 2,378 |
| 1980 | 2,953 |
| 1982 | 3,586 |
| 1984 | 4,397 |
| 1986 | 4,634 |
| 1988 | 4,962 |

This pattern has continued since 1988 and is repeated in all the countries of the affluent West (EU, USA, Canada, Australia, New Zealand and Japan). What social changes help to explain the massive increases in fresh food consumption? This pattern of consumption is associated with modern capitalist societies classified as 'postmodern'. The specifics of these societies are debatable, but of relevance here is that a high percentage of their populations are a relatively privileged, high-income, highly educated, well-trained professional and managerial class. This class enjoys large disposable incomes and is happy to spend some of this on fresh foods. The fresh produce is highly priced and increasingly it is cut, cored, chopped, sliced, diced, washed, mixed, colour-coordinated, in containers, attractively presented and ready to eat. These value-added processes (washing, cutting, etc.) greatly increase the profit margins in these goods, a fact reflected in the space devoted to fresh fruit and vegetables, in various states of preparedness, in supermarkets. In all Western societies there also exists a significant market of less privileged, less educated consumers, who have a limited disposable income and who eat mass-produced food, generally less healthy in type.

   A fundamental economic change facilitating the new regime has been the mobility of capital. Entrepreneurs from the West are investing in new

Box 3.2 (*continued*)

production locations in the developing world, where costs are low, especially labour costs, and good returns on capital are possible. Consider the following basket of goods and the geography of diets that they reflect: Kenyan mangetout, Zimbabwean and Jordanian green beans, Malaysian carambola, Egyptian garlic, Colombian apples and bananas, Zambian baby corn, Brazilian papayas. Other political and economic changes which have facilitated the increases in fresh-food marketing are the emergence of large integrated markets, such as the European Union and the North American Free Trade Area. These large regional groupings have large affluent markets for high-volume, high-value sales. These regional trading blocs have diverse climatic zones and varied labour markets, both of which are exploited to produce a diversity of foods, at all times of year.

One of the essential and obvious technological innovations that has facilitated the new food system is the establishment of global cool chains that integrate production with consumption. Fresh crops are chilled immediately upon harvesting, are stored and/or transported in controlled chilled conditions, delivered to chilled shelves in supermarkets, purchased by consumers and promptly stored in their refrigerators. Less obvious but other vital technological revolutions, essential prerequisites for the new food regime, are in information technology, retailing (large supermarkets) and marketing (niche marketing).

sell food products found across the globe, in the appearance of ethnic restaurants in the cities of the West (Balti in Birmingham, Nepalese in New York and Vietnamese in Vienna), and in food marketing strategies. There are other important aspects of this globalisation process, less obvious, but vitally important. Numerous themes could be examined (capital flows, labour regimes, export/import analysis, marketing strategies) but I have selected two major issues that illustrate this general process of globalisation. These issues are the role of transnational corporations (TNCs) in the global food system, and the diffusion of Western agricultural farming practices. These are not distinct or separate trends, as will become clear; they are parallel and intricately related processes.

**Plate 3.1** Exotic foods

**Plates 3.1–3.3** The globalisation of affluent diets

'Exotic' foods from across the world are found on supermarket shelves in the affluent world. Consumers are encouraged to try new and different foods. These goods have high profit margins (Plate 3.1). Highly processed, ready-made sauces available in the North allow people to become instant 'global cooks'. Thai, Indian, Mexican and Chinese are most popular in the UK, although there are at least twenty other 'ethnic' foods now available (Plate 3.2). Choice characterises affluent diets – deciding which apple, tomato, potato, lettuce, rice type, etc. to buy can be baffling (Plate 3.3).

*Photos*: The author.

Plate 3.2  We're all global now

Plate 3.3  What sort of apple did that recipe recommend?

## TNCs and the global food system

> While documenting the dominance of a handful of conglomerates in the
> US food system, we constantly encountered international activities of the
> same firms.
>
> (Heffernan and Constance, 1994, 40)

By the middle of the 1980s research on the globalisation of the agricultural and
food system was revealing the power of the TNCs. Whether in the production
of farm inputs, the trading of commodities or in food processing, there is a
marked concentration of ownership and control (matched at national and
regional levels by increased dominance by a few major retailers). These TNCs
have assets at their disposal which dwarf the annual budgets of some
developing states and information systems which most nation-states would
envy. Developments in information systems technology has been essential to
their gargantuan nature; electronic mail, satellite weather information and their
privileged access to new technologies both confirm and assure their dominance.
Combined with the BWIs they are vital agents in the transformation of the
global food networks.

   These conglomerates have diverse operations in numerous locations across
the globe. One of the largest food transnationals in the world is Cargill: with
its headquarters in Minneapolis, Minnesota, it is the world's largest grain
trading firm. It has operations in 49 countries with more than 800 offices and/or
plants employing more than 55,000 people. It trades in 103 commodities,
ranging from apple juice, wool, corn, oats, rice, sunflower meal, citrus pulp
and cocoa butter to ferrous metals, gold, platinum, salt, various fuel oils and
gases such as nitrogen; its other activities include ocean freight, financial
instruments and investor services.

   A sample of its activities in 1990 indicates its global character: it acquired
United Agricultural Merchants from Unilever; created, with Nippon Meat
Packers of Japan, Sun Valley Thailand as a joint venture; built a $1.25 million
molasses plant in the USA; bought Alexander, a New York feedmill; agreed
to buy for its subsidiary, Excel, Emge Packing of Indiana; announced plans to
open offices in Moscow and Warsaw; expanded its corn milling plants in
Tennessee; and planned an $8.3 million expansion of its poultry processing
plant in Florida. Cargill also funded research in plant and animal breeding. The
financial marketing department traded in financial instruments and foreign
currencies for Cargill and other companies and it provided commodity and
financial futures brokerage services and consultancies. Its stated aim is to

double its size every five to seven years (Tansey and Worsley, 1995; Heffernan and Constance, 1994, pp. 29–51).

Transnationals, by their nature, tend to engage in business with each other. Cargill has a joint venture with Nippon Meat Packers, Japan's largest meat packer, to produce and process broilers in Thailand for export to Japan. In this arrangement, Cargill provides vertically integrated production and Nippon provides access to Japanese markets. Mitsubishi has an operation in Brazil for export to Japan, and Mitsui has a similar operation in Malaysia, as does Ajinomoto in Brazil. Other recent examples suggest the magnitude of these connections. Tyson Foods (USA) has a joint-venture agreement with C. Itoh of Japan to produce broiler chickens in Mexico for consumption in Mexico, the USA and Japan. Tyson also ships chicken legs from chickens reared in the USA to Mexico for further processing for the Japanese market. Breast meat is removed for use in the US fast-food industry, then the remaining leg quarters are deboned in Mexico, where labour is cheaper: $4 a day in Mexico and $40 in the USA. Thus we have a variant on the global car concept – the global chicken nugget. The evidence supports the hypothesis that the TNCs are creating a global agro-food complex based on global sourcing of inputs, sites and output markets (Heffernan and Constance, 1994).

TNCs are a major driving force behind the establishment of a global agri-food complex based on global sourcing, and sophisticated technology is applied at every level of production, marketing and distribution. Global sourcing means that the decision-makers in TNCs can evaluate the merits of myriad locations for agricultural production locations and processing facilities. The advantages of global sourcing for TNCs include the ability to negotiate favourable terms (tax incentives, wages, labour contracts, etc.) by playing different potential locations against each other. For example, if a TNC wants to expand banana production it can negotiate with numerous Central American and Caribbean countries as well as the Philippines, all eager to attract foreign investment and employment. These countries will vie to host the TNC and its new operations.

Globally dominant TNCs are based in the USA, Japan and the EU, while some regionally significant ones are based in the newly industrialising countries. Increasingly, TNCs have joint ventures; so, for example, a US-based company will have a joint venture with an Italian one to have an entrance into the EU, or a Japanese one to exploit opportunities in East Asia. Table 3.1a and b show that their vision of a global agri-food complex is becoming a reality. TNCs have truly global ambitions, unbridled by answering to any nationally based constituency.

**Table 3.1** The power of the transnationals: (a) top ten pesticide suppliers in 1988; (b) top ten seed companies in 1988

(a)

| Company | Pesticide sales (US$ billion) | % of global market |
| --- | --- | --- |
| Ciba-Geigy (Switzerland) | 2.14 | 10.70 |
| Bayer (Germany) | 2.07 | 10.37 |
| ICI (UK) | 1.96 | 9.80 |
| Rhône-Poulenc (France) | 1.63 | 8.17 |
| Du Pont (USA) | 1.44 | 7.19 |
| Dow Elanco (USA) | 1.42 | 7.11 |
| Monsanto (USA) | 1.38 | 6.89 |
| Hoechst (Germany) | 1.02 | 5.12 |
| BASF (Germany) | 1.00 | 5.00 |
| Shell (Netherlands/UK) | 0.94 | 4.69 |
| **Total** | **15.00** | **75.04** |

(b)

| Company | Seed sales (US$ millions) | % of global market |
| --- | --- | --- |
| Pioneer Hi-Bred (USA) | 735 | 4.90 |
| Sandoz (Switzerland) | 507 | 3.38 |
| Limagrain (France) | 370 | 2.46 |
| Upjohn (USA) | 280 | 1.87 |
| Aritrois (France) | 257 | 1.71 |
| ICI (UK) | 250 | 1.67 |
| Cargill (USA) | 230 | 1.53 |
| Shell (Netherlands/UK) | 200 | 1.33 |
| Dekalb-Pfizer (USA) | 174 | 1.16 |
| Ciba-Geigy (Switzerland) | 150 | 1.00 |
| **Total** | **3,153** | **21.01** |

*Source*: Hobbelink, H. (1991) *Biotechnology and the Future of World Agriculture*, London: Zed Books, pp. 44 and 46. Also Tables 4.4 and 4.6 in Tansey and Worsley, p. 45

## The diffusion of Western agricultural practices

TNCs are vitally involved in promoting industrial agriculture, the basis of the modern agri-food system, in the developing world. Beginning in the 1930s in the USA, the character of agricultural production in the West changed in a variety of important respects. One of the most important was that connections between the industrial and agricultural sectors intensified. Previously farming had been dependent upon the weather and on-farm inputs but industrial agriculture is now completely dependent upon inputs from the industrial sector: farm equipment and fuel to drive it, seeds and feeds, drugs required to maintain animal health when intensive farming is the norm, and chemicals. Dependence on products from the petrochemical industries is most notable; fertilisers, pesticides and herbicides are all based on the petrochemical industry.

One of the most important issues which has emerged in response to the diffusion of this type of agricultural production is its environmental implications and its sustainability. These are pressing questions, allied to the topic of this text, but their importance and complexity warrant fuller discussion than space here allows. I draw attention to some of the issues in Box 3.3. Central to my concern in this text is to investigate how this dependence alters the entitlement context at the international and national level. Analysis of two basic questions will help our understanding of world hunger: what are the international ramifications of this dependence for the countries in the developing world and how does the promotion of this type of farming influence entitlements of producers and consumers in the developing world? I examine the first question below, the second question in the next chapter.

**Box 3.3**

### Industrial agriculture and the environment

Since the 1950s, agricultural production in the developed world has been transformed. The main characteristics of the new production system are high-energy inputs (fuel to power heavy equipment, fertilisers, pesticides, herbicides produced from petroleum), intensive animal husbandry and crop production (only possible by maintaining high applications of medicines/chemicals), monoculture (repeated production of one animal product/crop in the same building/field, which requires heavy applications of chemicals/medicines), and specialist production to rigid specifications demanded by food processors and retailers (controlled temperature, water and light inputs, growth regulators, growth stimulants, hormone injections, etc.).

Increasingly, this type of production is being regarded as unethical (animal rights activists at one extreme and many less extreme but still concerned campaigners; note the demonstrations against the export of calves to the Continent from Britain recently), environmentally disastrous and unsustainable (ground and surface water pollution, soil erosion, chemical residues, heavy water and fuel requirements). In the developed world these concerns are reflected in the emergence and success of many groups which argue against such production methods. In the developing world resistance to these methods of food production exists too. In addition, concern is voiced about the fact that the production and use of chemicals occurs in circumstances where governing legislation is weak and implementation even weaker. In some ways the BSE scare exemplifies some of these concerns.

*(continued on p. 58)*

**Plate 3.4** Agricultural machinery, Kansas, USA

**Plate 3.5** Hybrid maize varieties, Kansas, USA

**Plate 3.6** The Hoover Dam, Colorado, USA

**Plates 3.4–3.6** Agri-business

Agricultural production in the North depends on massive capital inputs; examples of machinery, scientific research and irrigation are illustrated above and opposite. This type of farming is being exported to the South very rapidly and Newly Agricultural Economies are emerging; already Thailand, Brazil, South Korea, Mexico and the Philippines among others have emulated Northern methods of intensive agricultural production. Debates about the sustainability of this type of agriculture are intensifying as ground water is depleted and/or polluted, soil erosion continues and concern about pesticide and herbicide residues increases.

*Photos*: The author.

Box 3.3 (*continued*)

### 'It's not "natural" for cows to be cannibals'

This is the sort of headline the BSE scare launched at its height in spring 1996, and it reflects concerns about several things. Anxieties exist about the health of the modern food system; previous 'food scares' and the most recent about sheep, the 'theoretical and hypothetical' possibility that they too can get BSE, all indicate that public confidence in contemporary farming methods is fragile. They also reflect the suspicion held by some that there was collusion between government, its health officials and the farming and agri-business lobby to manipulate the information released to the public. Certainly fears were aired that this issue pitted 'big business' against the consumer and that all the power was in the hands of the former. Increasingly some consumers feel alienated from food production, processing and supervisory systems. Specifically the headline reflects consumer unease about industrial food production, in this instance the protein-enriched animal feeds supplied by the agricultural conglomerates to farmers in Britain.

For British agricultural interests (farmers, abattoirs, processors, packers and distributors) the crisis is deep because their market is global and it has collapsed. In addition, it is evidence that technological interventions in food production can go very wrong and today, with standardisation and mass-production in food production systems, one mistake can have serious ramifications across large integrated economies, in this case across the EU. It is interesting to consider the role of the media too, especially in this case, where cows were seen staggering about on people's TV screens. The 'scare' certainly initiated discussion of food production methods and perhaps won more converts to vegetarianism and/or organic farming.

As the character of farm production has changed, the power of the TNCs has increased; these supply many of the vital inputs for industrial farming. A few large TNCs tend to dominate the production, distribution and hence type and cost of farm inputs; their influence over the direction of biotechnological research is typical of their power. Pricing policies of the companies and their control over farm inputs mean that the decision-making powers of governments and farmers, anywhere in the world, are curtailed. It is reasonable to expect that the power of the TNCs will intensify as our diets become more global and it is, then, pertinent to ask, 'Given that TNCs increasingly source their products from "have-not" nations and sell their products to "have" nations, will this trend in global food concentration just continue the long-term inequalities in relationships between rich and poor nations?' (Heffernan and Constance, 1994, 48). Evidence presented in the next chapter suggests that it will.

## Conclusion

This chapter illustrates how interconnected the global food system is at present; that is the global context within which we can begin to understand people's entitlements. Globalisation trends are undeniable but in conclusion it is important to draw attention to some of the limits of a global perspective. Anyone who has travelled in the South, especially outside the major cities, will be aware of how very *local* many diets remain. Most people in the South do not have sophisticated supermarkets with refrigeration facilities, or access to motorised transport on which they depend, and their homes have not got fridges or freezers – all vital technologies for Western shopping and eating habits. Most of the poor must spend time every day, or every two days, shopping for fresh foods, which are consumed within 24–48 hours. This helps to explain the prevalence and wonderful variety of dried foods (lentils, beans, rice, herbs, etc.) which form the staple goods in local markets and shops, and also the importance of vegetable growing and small-animal husbandry on domestic plots in urban and rural areas. This chapter emphasises the processes of globalisation which are influencing agricultural and dietary changes in the North and South. It is not meant to suggest that these processes are all-encompassing, or that all national, regional and local factors have been obliterated. Factors at all these levels are the subject of the next few chapters.

## Key ideas

1 A number of factors at the international level govern access to food. Particularly important are patterns of debt and capital flows, the policies of the International Monetary Fund and World Bank, and the behaviour of transnationals.
2 Historical legacies continue to shape the pattern of food availability in the world; trading patterns, property relations and access to technology are among the most important.

**Plate 3.7**  Local millet porridge, northern Nigeria

**Plates 3.7–3.10**  The persistence of the local

While global processes are undoubtedly influencing food entitlements in the South, and competition is eroding local subsistence production, many diets remain predominantly local in character. The implements as well as the food in Plate 3.7 are locally produced. Meat (Plate 3.8) is bought and consumed quickly in northern Nigeria because refrigeration is rare. Fishing on the River Sokoto is uncapitalised and the catch is consumed locally. Sugar cane is sold in small pieces as a snack in northern Nigeria. Diets in the South are being transformed, however, and evidence of transnational marketing is everywhere and increasingly effective.

*Photos*: Dr Hamish Main, Staffordshire University.

**Plate 3.8**  An Islamic butcher, Kano, Nigeria

**Plate 3.9** Fishing in the River Sokoto, Nigeria

**Plate 3.10** A sugar cane seller, northern Nigeria

3 Changes in the diets of people in the North have implications for the diets of people in the South.
4 Complex patterns of production, processing and marketing link the elements of the global food system. The globalisation of the food system is intensifying and transnationals are a principal force behind these changes.
5 Food production is being transformed in the South as intensive, high-input agriculture, associated with the North, replaces traditional methods of farming.
6 In the global supermarket, food is drawn to the affluent markets of the North and away from poorer markets in the South.

# Kenya

## ROAST CHICKEN WITH PEANUT SAUCE                              Serves 4–6

Coffee, tea, fruit and vegetables – mostly grown around Mount Kenya – are some of the country's prime exports. Falling prices for tea and coffee have meant less foreign exchange to buy oil and other necessities.

In this recipe the peanut sauce is handed round separately. Serve the chicken with sweet potatoes or rice.

Jane Nash, Ashford, UK

## INGREDIENTS
3 lbs/1.5kg chicken
3 tablespoons/40g butter, margarine or ghee
1 cup/240ml milk
2 medium onions, finely chopped
salt and pepper
1 tomato, finely chopped
½ cup/110g crunchy peanut butter

## METHOD
1 Heat the oven to 325°F/160°C/Gas 3.
2 Sprinkle some salt and pepper over the chicken and dot it with the margarine, keeping 1 tablespoon back for later use. Then bake it in the oven for about 1½–2 hours until done, spooning the margarine and juices over from time to time.
3 About half an hour before the chicken is ready, melt the remaining margarine in a saucepan and sauté the onions until golden. Stir in the chopped tomato and cook for 5 minutes.
4 Add the peanut butter and blend it well. Now gradually stir in the milk and seasoning. Then cover the saucepan and simmer for about 20–30 minutes, stirring occasionally. Pour this sauce into a dish and then serve the chicken, handing the sauce round separately.

# 4
# National perspectives

## Introduction

Chapter 1 emphasised that a central task of any government is to secure food for its population. It is interesting then to examine why in the late twentieth century some governments fail miserably to do so. This chapter examines a variety of factors which explain that failure. The emphasis of Chapter 3 was on factors at the international scale which prejudice a country's ability to command food, i.e., how international factors can undermine a country's ability to generate resources needed to command food for its population. This part of our analysis explores factors at the national level which help explain why, within countries, entitlements vary. Some of the myriad ways governments interact with local food security are shown in Figure 4.1.

It is often difficult to discern the boundaries between international and national decision-making; international actors often form alliances with nationally potent actors and the context of national decision-making is structured by decisions made elsewhere. However, the distinction is useful because it helps to identify the variety of agents and actors which govern access to entitlements and food. Generalisations about national policy are necessary but it is important to emphasise that circumstances in the developing world are extraordinarily varied and dynamic:

> In some developing countries such as Mexico and Brazil, the quantity of food available is well above requirements (in fact, food availability per person in Mexico is estimated by FAO to be almost the same as in many

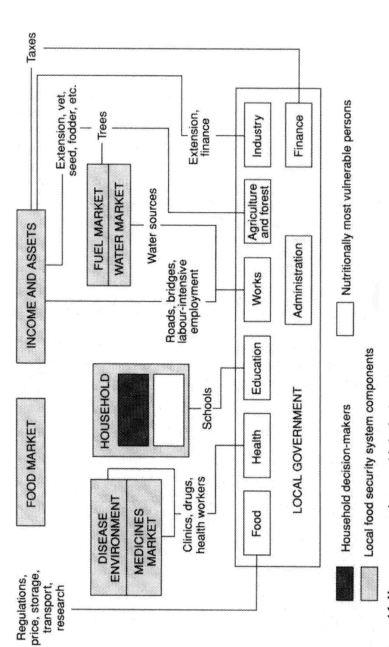

**Figure 4.1** How governments interact with the food security system

*Source:* Hubbard (1995, 10)

industrialised countries of the North). None the less, rates of undernutrition
and poverty in these countries are as high as in other developing countries
such as Nicaragua or Senegal with barely enough to meet their theoretical
consumption needs.

(Barraclough, 1991, 2)

## Historical legacies

European colonisation occasioned numerous changes in the developing world
which continue to shape people's entitlements; some people benefited from
these changes and others suffered a decline in their circumstances. Within
colonial states, access to productive resources, land, mineral wealth, etc., were
always skewed, and patterns of inequality established in the past persist. In
general, a few people controlled the majority of the resources and the great
majority were in control of very few resources. The case of property relations,
the rules that govern ownership and access to land, is an instructive example.

Control over land was an important element of colonial power and patterns
of land ownership changed with colonialisation to benefit some groups and
marginalise others. People who own or have access to large landholdings seldom
want for food, but some of the most malnourished populations in the contemporary world are peasants who have small plots or landless labourers who
have none. Inequitable land ownership certainly helps to explain the contrasts
in entitlement to wealth and food in many countries in the developing world.

In Latin America, large land-owning interests have been dominant in all
powerful institutions since Spanish and Portuguese colonialisation (Bernstein,
Crow and Johnson, 1992). In the Philippines too, Spanish land-owning
structures were introduced that prevail up to the present. The land question
fuelled the independence movement in India, although land reforms after
independence were limited in their impact. In Africa, north or south of the
Sahara, access to land and its utilisation was radically transformed in the
nineteenth century. Throughout the developing world inequalities in access to
this essential resource continue to have serious implications for the distribution
of entitlements, and many experts maintain that while it exists poverty and
hunger will endure.

Land reform means land to the tiller – the redistribution of the rights
associated with landownership in favour of those physically cultivating it, at
the expense of landlords, speculators, and other landholders. The beneficiaries are precisely those groups with the most inadequate access to food.

(Barraclough, 1991, 102)

Land reform is difficult to institute, because the land-owning élite can usually obstruct or dilute proposed changes. The Philippines, all of Latin America except Cuba, and the countries in South Asia and North Africa all offer examples of limited land reform because of the power of landed interests. The impact of land reform efforts has been very mixed and depends on the commitment and capacity of the state, not only to implement them, but to provide infrastructural support to ensure their success. If land reform is implemented effectively, the beneficiaries enjoy improved food security. In China, a peasant-led revolution transformed control over land, and food security was enhanced for the majority (Box 4.1). Japan, Taiwan and South Korea all experienced externally imposed land reform in the twentieth century and have since evolved relatively egalitarian peasant farming sectors. Their success, however, occurred in unique circumstances after the Second World War, when the West was preoccupied by containing Communism and there was the simultaneous expansion of manufacturing, which absorbed rural unemployment. Specific geopolitical circumstances help to explain these successes and are unlikely to be replicated.

Box 4.1

### China: a conflicting picture

It is instructive to examine the case of China and food security since the revolution of 1949 for several reasons: before 1949 famine and malnutrition were endemic; since 1949, with the exception of the famine of 1959–61, famines have not occurred and malnutrition has declined despite rapid population growth, from 500 million to over 1,183.6 million (1992). China has about the same amount of cultivated land per capita as Bangladesh but less than India. Its physiological population density is, therefore, higher than in these two countries, where hunger is still endemic. Until the reforms of 1978, development strategies were based on self-reliance and state control; the changes since 1978 contrast with previous policy and are having debatable consequences. In addition, all the issues that this text claims are relevant to the global debate about hunger can be posed of the Chinese case.

Before the 1949 Communist revolution the food security situation in China was drastic. Between 1900 and 1949, the Chinese had suffered foreign invasion, civil wars, recurrent floods and droughts; all of these and the extremely inequitable property relations which prevailed meant

Box 4.1 (*continued*)

that food security was the exception rather than the rule. In rural areas, land ownership was inequitable, landlessness was high, tenancies were insecure, rents were high and food supplies precarious. Things were no better in urban areas where unemployment was high and wages were low. Development strategies have varied and are contentious within and outside China, but with the exception of the famine years 1959–61, the general nutritional status of its expanding population has improved; malnutrition has not been eradicated and regional and social differences are still serious, but access to a better diet is a reality for more of its population than previously. This success reflects the political commitment of the state and its prioritising of food security.

The revolution in China in 1949 was based on the transformation of social relations, and the emphasis between 1949 and 1976 was on collective, communal solutions to the problems of production and consumption. The traditional central role of the family in welfare issues was eroded. Strategies between 1949 and 1978 included radical land reform, state control of cereal prices and distribution, rationing and securing subsidised food for urban populations, the promotion of rural industrialisation and employment, and population policies. During this period, the guiding principle was that the state would determine patterns of production and distribution. Land was held and worked communally, and each unit (commune) was required to provide the state with quotas of food grains. This system worked to provide more food security than peasants had previously enjoyed, but by 1978 the system was failing in several respects. Crop yields particularly were stagnant or falling, so in a major change of course the state initiated radical changes in all sectors of the economy, including the massive agricultural sector, which still absorbs the bulk of Chinese labour.

The changes ushered in since 1978 have challenged the doctrines which had previously seemed so secure: centralisation, communal ownership and egalitarianism. Decentralisation, privatisation and reward for effort have become the new guiding principles. There is a new emphasis on cash crop production and regional specialisation is encouraged. At the centre of these changes is the role of the peasant household: after an interlude of some thirty years, it has re-emerged as the primary unit through which economic and social relations are based.

Box 4.1 (*continued*)

While the peasant household was probably never completely replaced, its role was certainly viewed as secondary to the role of the collective between 1949 and 1978. Reforms initiated since 1978 have altered that emphasis; the peasant family is once more at centre stage. The peasant household has virtually replaced the collective as the unit of production, consumption and welfare. Households are encouraged to maximise their productivity, to market that produce and to retain the profits. There is evidence that these changes have stimulated agricultural production and wealth creation in the countryside. There is cause for concern, however.

Some observers have noted increased levels of regional and social polarisation. As some of the most viable household units, in the most favoured regions, have capitalised on the opportunities offered, those less viable units in less favoured regions have lost the security provided by the pre-1978 regime. Markets for food are thriving, but, as argued elsewhere in this text, these do not provision individuals or groups without entitlements. Some experts express anxiety about the impacts of the new policies on population growth rates. As household labour means increases in productive capacity, new incentives to have children are appearing.

This brief overview of the Chinese case reflects the importance of national policy decisions on food security at all the levels explored in this text. National policy is mediated by a range of other variables: class, gender, ethnicity and age at every scale of analysis, from the major regional administrative units in China down to the household level. A glimpse into just one household is instructive. Although referring to contrasts within one household, the following image is suggestive of the conflicts and contrasts facing China more generally since the reforms of 1978. It is of:

. . . a single household with two rooms, one to each side of a central kitchen. Visitors were received into one or other room and between the two (heaven and earth?) lay the centre – the hearth of the household. The one to the left, where I stayed with other female members of the household, was meagrely furnished and more than half taken up with a kang, a large platform bed heated by a fire within and on which the colourful quilts of the two daughters of the household

Box 4.1 (*continued*)

> lay carefully folded. The walls were adorned with portraits of
> revolutionary leaders including Mao Zedong, a number of political
> posters and framed Communist Party certificates honouring the 'good'
> Party and community services of the cadre mother of the family. The
> other, to the right, where I lived by day could not have provided a
> greater contrast. It featured a bed, a comfortable sofa and a substantial
> marble-topped coffee table made by the local village factory of which
> the father of the family was the manager. Prominently displayed was
> a ghetto blaster, a colour television, a video-recorder and a bright green
> fridge-freezer. On the walls were a large glossy calendar displaying a
> swimsuited female figure and a musical clock whose closeted bird sang
> 'Happy Birthday' every hour!
>
> (Croll, 1994, 225)
>
> *Source*: Based on Barraclough (1991) and Croll (1994).

Most former colonial states have not seen transformations of property
relations; more frequently a national élite assumed power with independence.
In many cases the politicians were educated in the capitals of the colonial
powers, and their wealth, education and culture meant that they had little in
common with the majority, mostly rural, inhabitants of the countries they
governed. This had several consequences for the policies they adopted. With
a few exceptions they were an urban-based élite, who accepted the development
philosophies which prevailed in the West and which were promulgated by the
Bretton Woods institutions (BWIs) and the armies of experts sent from the
North to 'develop' the South. These development philosophies and their
implications for agricultural policy and food security are examined next.

## Development strategies

Between the 1950s and the 1980s, the *modernisation theory of development*
influenced policy-making in the developing world. Adherence to this ideology
continues to have important ramifications for the distribution of wealth, power
and food in the developing world. This theory held that the 'path to progress'
was through industrialisation, so the majority of available state funds were

devoted to furthering that end. Because most inhabitants of the developing world are still rural, this policy effectively cut them off from state investments. State funds were funnelled towards industrial activity and urban infrastructure and development theorists argued that the wealth created would 'trickle down' eventually to the poorest sectors of society. In fact, the outcome was the opposite and in very many places social and economic polarisation increased. There was an associated exacerbating factor: state funds employed to promote industrialisation were often derived from taxing the rural sector. Even China, which since its revolution in 1949 purported to follow a different 'path' and prioritise its peasant class, in fact also taxed the rural poor to fund urban and industrial infrastructural improvements. The majority of the poor in the developing world are still rural dwellers, although that situation is changing (Table 4.1), and the lack of infrastructural improvements is reflected in the disparities between rural and urban socio-economic indicators.

Large-scale dam construction is one of the best symbols of the dominance of this approach. Massive state investments were devoted to grandiose schemes, often with the assistance and encouragement of international institutions like the IMF and WB, which subsidised industrial development or, if associated with irrigation projects, increased production of commercial

**Table 4.1** Growing urbanisation in selected countries, 1992

| Country | % urban | | |
|---|---|---|---|
| | 1960 | 1992 | 2000* |
| Malaysia | 27 | 51 | 57 |
| Brazil | 45 | 76 | 81 |
| Bangladesh | 5 | 17 | 21 |
| South Africa | 47 | 50 | 53 |
| Sri Lanka | 18 | 22 | 24 |
| Indonesia | 15 | 33 | 40 |
| Egypt | 38 | 44 | 46 |
| China | 19 | 28 | 35 |
| Pakistan | 22 | 33 | 38 |
| India | 18 | 26 | 29 |
| Nigeria | 14 | 37 | 43 |

*Source*: Human Development Report (1995), pp. 184–5
*Note*: * projected.

crops for export. A rash of these schemes occurred in the 1960s: the Aswan Dam on the Nile, the Kariba Dam on the Zambezi, the Akosombo Dam on the Volta and the Kainji Dam on the Niger. There is now a vast literature which considers the severe negative social and environmental consequences of these projects. These schemes are testament to the prevalence of modernisation theories, and their impacts exemplify the shortcomings. Modernisation theories are faltering but are not dead, because they often benefit powerful economic interests (Elliott, 1994, 47–9): one of the most recent contentious dam projects is the Narmada project in India. The controversy over dam construction in Burma also illustrates the longevity of the modernisation philosophy and, if pursued, it will exemplify the conflicts of interest, national and international, that are implicated in the policy-making process.

Enthusiasm about the green revolution dating from the 1960s is allied to development philosophies which prioritise production. The impact of the new technologies has been very diverse, depending upon the pre-existing social context. Certainly, some countries are now self-sufficient in cereal production and therefore less vulnerable in some ways to international price changes, but in other ways they may have become more exposed to shifts in international markets and dependence upon TNCs. As these new technologies required capital investments in irrigation, hybrid seeds, pesticides, fertilisers and herbicides, farmers who had larger farms and incomes could afford to invest in the new technologies, while peasant farmers with few assets were denied the opportunity. Thus the already relatively privileged became more so. Other negative impacts for the poorest sections of the rural population were increases in rents and the concentration of land ownership. In sum, increases in the production of food are not necessarily associated with its more equal consumption.

A still pervasive development philosophy is that developing countries can increase their earnings, and hence their national entitlements, by increasing their exports, especially of agricultural commodities. Heavy reliance on any one export makes a country's revenues vulnerable to many unpredictable variables, including world price fluctuations, changes in geopolitics, exchange rates, changing technologies and world recession.

The effect on food security of concentrating on cash crops for export is ambiguous and depends upon the wider social context of development policies. Certainly, in some situations it can have negative impacts for some of the poorest rural inhabitants. The examples of Honduras and Burkina Faso illustrate some of the negative implications of cash crop expansion when promoted in communities where access to wealth and political power are grossly inequitable (Box 4.2). Expansion of cash crop production may be beneficial for food security if land ownership is relatively equitable and if exports are diverse, but clearly a rush to promote cash crops for export may just intensify food insecurity, increase economic and political marginalisation, and increase the likelihood of environmental catastrophe. The beneficiaries of the expansion of Honduran shrimp farming are the consumers in North America

Box 4.2

## Cash crops and entitlements: some dangers

Honduras

Development strategies based on cash crop production for export have guided state policy in Honduras since the 1950s; the social and environmental impacts have been disastrous. This case also exemplifies the relationships between national and international factors.

Anxious about poverty and associated unrest in Central America, the United States launched the Alliance for Progress initiative in the early 1960s. The development effort received initial funds from the United States, which were complemented by multinational and national capital. It was designed to promote exports from Central America to the USA; special trading arrangements helped to encourage exports. The most important exports promoted from Honduras between 1960 and 1980 were cotton, sugar and beef; the importance of each varied in accord with the world market. The expansion of land devoted to all three products had serious social consequences for some of the poorest peasants as their entitlements were eroded: inequalities in access to land increased; many poor peasants were displaced to marginal lands. The environmental costs were serious too. Cotton cultivation depends on the heavy application of chemicals, which were often applied carelessly and without regard to environmental contexts. The expansion of cattle lands was associated, throughout Central America, with the reallocation of land from forest, fallow or food crops to pasture. In the region which experienced most of this 'development', malnutrition was widespread; a survey in 1982–83 found that 65 per cent of children under sixty months of age were stunted and that there was a close relationship between access to land and nutritional status.

During the 1980s, economic crisis was again threatening Honduras and the rest of Central America. Another initiative (1983), promoted and funded by the United States and multinational investors, was launched under the banner of the Caribbean Basin Initiative (CBI). The strategy was essentially the same as previously, the expansion of exports, although the goods had changed. The 'new' crops included fruits and lobsters but most prominent was the expansion of shrimp exports. Shrimp farming in the coastal regions of southern Honduras grew rapidly so that

Box 4.2 (*continued*)

by 1989, nineteen large shrimp farms and thirty-five smaller ones were raising shrimps for export to the cold buffet tables of North America. The social and environmental consequences of this expansion replicate all the most negative aspects associated with the earlier phase of agricultural transformation. Benefits have accrued to a small minority of local and international interests while local coastal livelihoods are threatened with complete collapse as the mangrove swamps upon which the coastal ecosystem depends are decimated. This assessment does not mean that the promotion of export agriculture is necessarily associated with an erosion of entitlements among the poorest sections of society and environmental degradation. The lesson lies in understanding how policies may be modified so that the exploitation of natural resources for export crops can both reduce poverty and be environmentally benign and sustainable.

*Source*: Stonich (1992).

### Burkina Faso

The expansion of large-scale sugar production during the 1970s produced a local livelihood crisis and a minor ecological disaster. The Comoe Sugar Company, a state-controlled enterprise affiliated to a transnational, obtained a concession for 10,000 hectares of good river bottom agricultural land to produce sugar cane. Peasants in the area were simply ousted, their villages and orchards destroyed. They were moved to nearby but inferior lands that were part of the fallow rotations of other peasant communities. Indemnity payments were grossly inadequate. When the company opened the sugar mill none of the former inhabitants was offered employment. Their new village lacked infrastructure, and crop yields from their new lands were lower than previously. The mill polluted the river downstream and sugar cane irrigation sucked so much water from above the mill that peasants there had to abandon irrigated vegetable and rice production. In 1984, a nutritional survey revealed that most peasants in the area were far below what they had been a decade earlier.

*Source*: Barraclough (1991, 32).

and Europe, who enjoy cheaper shrimps unaware of how their production is prejudicing the lives and livelihoods of the poor elsewhere. It is important that the real costs, social and environmental, of these cheap foods are appreciated.

Zimbabwe has diverged from the general pattern by concentrating on food production and the small farmer sector. A consideration of this effort to boost food security illustrates the problems faced by governments when they adopt these policies. Until recently, agricultural policies followed in Zimbabwe since independence in 1980 were viewed as relatively successful, boosting domestic food self-sufficiency and regional food security. The government decided to focus on the small farm sector and ended the previous policies, which had favoured the large-scale, predominantly white, commercial farmers. Instead it instituted pricing, marketing and extension services that favoured the small farmer class, who produced the basic food staple, maize. In 1980, this sector accounted for 5 per cent of all marketed grain surpluses; in the early 1990s this figure varied between 45 and 60 per cent. Marketing of produce was undertaken by a state agency, the Grain Marketing Board (GMB), which facilitated the collection and transport of surplus. In the late 1980s and 1990s, this agency had to extend storage and marketing services to deal with increases in supply. An essential element of the government's policy was that some strategic grain supplies were kept to ensure food availability in the event of poor harvests.

Problems emerged as world prices fell and debt repayments soared. In 1991, at the urging of the World Bank, a structural adjustment programme (SAP) was implemented (see Chapter 3). Although the WB agreed that some stocks should be stored to ensure food security, it argued that financial reserves were more important than food reserves. The government reduced price supports and maize production declined at the same time as the GMB sold off the bulk of reserve stocks. By early 1992, stocks had declined to 300,000 tonnes just as drought occurred and production plummeted. Famine was averted only by massive imports of food, some in the form of food aid.

This case illustrates how the international and national are connected and how food security strategies may be thwarted. The debate continues about how much to liberalise agricultural policy and how these policies will impinge on food supplies and food security in the developing world. Many Zimbabweans believe that a completely hands-off policy by the state would be suicidal given their unpredictable climate; they argue that some measure of state intervention to ensure grain in times of crisis is essential. It is amazing how vehemently the benefits of a non-interventionist approach to food production are preached by the Northern countries, when they have supported domestic food production, blatantly and relentlessly, and continue to do so despite the rhetoric of the Uruguay Round. In 1995, the common agricultural policy (CAP) in the EU

**Plate 4.1** Female labourers in India

**Plate 4.2** Seamstresses in Kano city, northern Nigeria

*Photos*: Plate 4.1, Dr Jennifer Elliot, University of Brighton; Plates 4.2–4.4, Dr Hamish Main, Staffordshire University.

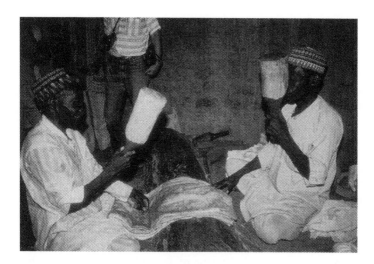

**Plate 4.3** Beating indigo dye, northern Nigeria

**Plate 4.4** Fuelwood for the urban market, northern Nigeria

**Plates 4.1–4.4** Diverse entitlements of urban and rural people in the South

People in the South have diverse entitlements; they rely on a great range of activities to survive. The socio-economic context of their livelihoods is dynamic; opportunities are constantly created and/or destroyed, with important implications for survival.

Consider the activities shown above and discuss some changes which could impact these sources of income.

received 46.9 per cent (35.6 billion ecus) from the total EU budget of 65 billion ecus. Change in the CAP will be slow and bitterly contested and offers an interesting contrast to the predicament of countries like Zimbabwe, whose efforts to protect food supplies and the small farming sector were challenged so hotly even as massive agricultural subsidies in the North continued.

While the role of development policy in general and agricultural policy in particular is obviously relevant to issues of food security, it is also important to recognise that the livelihoods of rural inhabitants are very diverse. The great majority depend on a variety of activities to ensure their survival. These may include remittances from family members working elsewhere, wages from occasional work in agriculture or industry, trading activities, craft production and numerous service activities. The great majority depend on a variety of activities to ensure their survival; these may include remittances from family members working elsewhere, wages from occasional work in agriculture or industry, trading activities, craft production and numerous service activities. For example, the introduction and diffusion of aquaculture in Africa may be locally significant for improving entitlements and diets, especially for women, who have often been marginalised in development programmes. Although African fish farming still only provides a small fraction of total fish production (in 1989 fish farming accounted for 68,000 of the total 3.8 million tonnes of fish produced in Africa), and is insignificant compared to the situation in Asia where intensive aquaculture has a long tradition, it is becoming increasingly important. Small-scale aquaculture in Africa doubled between 1980 and 1990 and the UN Food and Agriculture Organization estimates that it could be important in as many as thirty-one states in Africa – securing income from fish sales as well as improving the nutritional content of local diets. Strategies to reduce poverty therefore must include increasing employment opportunities and wealth-creating activities; such policies have had mixed results. A recent evaluation of Indian efforts to target public spending programmes to alleviate poverty concludes that these can only mitigate the harsh and polarising effects of more general economic forces, which are increasing the gap between the rich and poor (Ghosh and Bharadwaj, pp. 139–64, and Crow, pp. 251–73 in Bernstein, Crow and Johnson, 1992).

Urban populations in the developing world are growing both as a percentage of total population and in absolute numbers. Government policies structure the contexts within which these populations have access to a decent diet, just as they do for rural populations. The specific problem of urban food supplies has received less attention than the problem of rural food supplies, perhaps because the images of famine are often most serious in rural areas. Socio-economic statistics also suggest that urban populations are relatively privileged *vis-à-vis* rural populations. However, these data must be treated with caution because information about the poorest inhabitants in the major cities in the developing world is not always accurately recorded; recent migrants, highly mobile, and very poor populations may escape accounting efforts. Certainly, these statistics indicate that all is not well with the majority of the urban poor regarding nutrition and health. It is also important to stress that these populations are growing very rapidly, from natural increase as well as massive in-migration, so the question of food security for urban populations, already important, will become more so (see Chapter 3, food riots).

Drakakis-Smith (1994) has studied the impact of the world recession and SAPs on urban food supplies in Harare, Zimbabwe, in the 1990s. His analysis draws attention to the specific issues which surround urban food supplies and how the urban poor adapt to increasingly harsh realities imposed by external forces and implemented by the state. As discussed above, the government in Zimbabwe has since 1980 targeted programmes at populations usually neglected, but SAPs have required state spending on programmes for the poor to be cut. The urban poor may suffer most as the prices of food, fuel and education increase and their employment opportunities and wages fall. In these circumstances the urban poor are forced to devise novel ways to subsist. The Harare study reveals patterns that are undoubtedly repeated in very many cities in the South, with modifications which reflect specific local circumstances (Box 4.3).

Box 4.3

---

### Harare: the urban food system

As was emphasised in Chapter 3, certain global processes influence food systems everywhere – evidence of this is found in Harare. One characteristic is that the great majority of urban dwellers purchase the food they consume (rather than grow it). A second is that as far as marketing is concerned, there is a shift to more intensive, capitalised forms of supply and retailing so that 'in cities of such diverse size and locations as Hong Kong, Harare and Port Vila over half the food is now retailed through supermarkets' (Drakakis-Smith, 1994, 8). A third is the shift in diets away from traditional foods and towards Westernised commercial foods, i.e., convenience foods. Despite these general trends one of the most striking characteristics of Third World cities is the persistence of the street vendors of food. This informal sector persists in urban food distribution

*(continued on p. 82)*

**Plate 4.5** Urban street trader, Calcutta, India

**Plate 4.6** Street hawker, Dhaka, Bangladesh **Plate 4.7** Urban market, Dhaka, Bangladesh

*Photos*: Plate 4.5, Dr Jennifer Elliott, University of Brighton; Plates 4.6–4.8, Dr Sabiha Sayeed, Staffordshire University; Plate 4.9, Dr Hamish Main, Staffordshire University.

**Plate 4.8** Urban agriculture, Dhaka, Bangladesh

**Plate 4.9** Urban agriculture, Delhi, India

**Plates 4.5–4.9** Urban food provision

Intensive agriculture is common in many cities of the South and traders and hawkers are a vital element in urban food distribution systems. Ready-made snacks, prepared in the 'informal' sector, are sold to workers in the 'formal' sector; in some parts of the world the income from these sales is the most important form of entitlement.

**Plate 4.10**  Recent migrants to Delhi
Urban areas in the South are expanding rapidly because of natural increase and in-migration. Changes in agricultural production have caused massive out-migration from rural areas in all countries of the South; the majority of the world's poor will be urban dwellers by the early decades of the next century. The urban poor suffered throughout the 1980s from the impact of structural adjustment policies; wages fell and food prices increased (see Box 4.3).

## Box 4.3 (*continued*)

systems and may even be expanding among the poorest sections as SAPs exacerbate urban poverty; simply, food from street vendors is cheaper. Unfortunately, most officials in the developing cities frown on street vending and the activity often experiences severe criticism and repressive policies.

Another response to SAPs and increases in food prices is the expansion of urban food production. While it is probably true to understand the expansion of urban food production as a survival strategy by the poor, there is growing evidence to indicate that some of this food is produced by middle-income households and entrepreneurs. Urban food is grown

Box 4.3 (*continued*)

in private urban gardens in the less intensively developed towns and cities and on the urban periphery. Animals can be reared and crops cultivated in both of these places for consumption by the household, to barter or for cash; the particular combination of crops and consumption or exchange varies. Unfortunately, these innovative ways to establish a limited degree of food security are also frowned upon by officials concerned about the image of their cities; even in 1991, when the country was in the grip of a drought-induced food shortage, 'illegal' plots of food were slashed while the mosquito breeding grounds remained untouched! Drakakis-Smith concludes his paper with two pleas: more research into urban food systems, and immediate and more sympathetic national and local policies to support these efforts by urban food retailers and producers to survive in such unsympathetic macro-economic environments.

*Source*: Drakakis-Smith (1994).

## Conclusion

This chapter has considered some of the national processes which have structured the entitlements to food within countries in the developing world; the emphasis has been on inherited circumstances and the development philosophies which have been influential in directing social and economic change in the developing world since the 1950s. To summarise:

- those with access to land have fared better than those with none or very small plots;
- the process of increased commercialisation tends to increase land concentration;
- the emphasis on industrialisation and urbanisation has prioritised investment in those sectors of the economy;
- investment in the agricultural sector has favoured the cash crop sector rather than the food crops sector, i.e., cash crops for export rather than food for domestic consumption;
- the political clout of the organised, concentrated urban populations has been more effective in influencing governments and their policies than that of dispersed rural populations, although this is not to suggest that all urban dwellers enjoy adequate diets;

- there is a growing crisis of food provisioning in the rapidly expanding cities of the developing world.

The African National Congress in South Africa issued a policy document in 1994 which is worth considering in the light of the issues raised in this chapter, because it emphasises all those things that traditional policy has ignored; it remains to be seen if, and how, the rhetoric in this document is translated into effective policy:

1 The agricultural mission of the African National Congress is to achieve equitable access to and optimal use of agricultural resources to ensure:

  - affordable and sufficient food and fibre for all South Africans;
  - a life of dignity for all on the land;
  - sustainable rural development;
  - the creation of employment and the elimination of rural poverty;
  - just rewards for skills, energy and enterprise;
  - full realization of agriculture's contribution to economic development;
  - conservation of our natural resources for the benefit of future generations.

2 The ANC is committed to ending poverty and malnutrition and to ending policies which have failed to meet the nutritional needs of the majority. Emphasis will be placed on ensuring low and stable prices for basic foods of low-income consumers. The basic aim will be to ensure household food security – access for all people at all times to enough food for an active and healthy life. Food insecurity is most prevalent in rural areas, highlighting the need for improving production and income-generating activities. The inability of the majority of rural people to produce a marketed surplus, or even meet their subsistence needs, is a reflection of their limited access to land, water, credit and markets, and the failure of research and extension services to provide appropriate technologies.

3 The ANC will introduce measures to improve rights to land and access to credit and other resources to improve small-holder productivity and food security. Research and extension expenditures will be redirected towards improving technologies for labour-intensive production and on-farm storage. More generally, the ANC will devise a rural development strategy which maximises employment and income-generating activities in the rural economy as a whole.

4 To tackle poverty, correct historical imbalances and revitalise the rural economy, an ANC government will redirect spending towards the upliftment

of rural people. Funds for this reorientation could be found first by redirecting spending on the inequitable and inefficient subsidization of the large farm sector. Large-scale farming will continue to have a vital role, but its access to land will be reduced with land distribution to small-holders. It must learn to cope without government subsidies, and in an environment of full legal rights for farm workers.

5 The state must provide much of the funding for rural development but the setting of priorities and control of funds should devolve to local communities. Structures must be set up in such a way that different groups can lobby for different ways of using the funds available. In particular, the economic difficulties of rural households headed by women must be recognized.

(ANC Agricultural Policy, May 1994)

This chapter considered the class/rural/urban differences which influence policy and food entitlements but has completely neglected the gendered character of debate and policy. The next chapter is devoted to this issue.

## Key ideas

1 National policies in the countries of the South are very diverse and reflect their different colonial legacies and contemporary power structures.

2 One of the most important legacies of colonialism, which has a direct bearing on a population's ability to command food, is the system of inherited land-owning patterns. Rural landlessness is a serious problem in many countries in the South and is politically difficult to address because those in power often own the land. Entitlement to land is limited to the élite in many countries.

3 Development strategies which have been promoted since the 1950s have neglected the rural sector and favoured industrial and urban expansion; food production and food security have not had a high priority.

4 Increases in national food production are not automatically associated with reductions in hunger. Occasionally the reverse occurs, and increases in food production are accompanied by increases in the incidence of hunger as access to entitlements is eroded for the vulnerable (land becomes expensive and concentrated, common property resources are privatised, cash crops replace food staples, etc.).

5 The debt burden continues to drain resources from the South to the North and reduces the ability of Southern governments to enhance food security.

6 Although rural hunger is still a quantitatively larger problem than urban
hunger, the situation is changing rapidly, and unless the direction of change
is reversed, by the next century hunger will be a predominantly urban
problem.

# Jordan

## BASAL BADAWI (Onions with meat, nuts and fruit)                 Serves 4

This is a traditional dish of the Bedouin, the nomadic 'guardians of the desert'. It is easy to prepare and can be made festive with the addition of saffron or turmeric in the accompanying rice. Red lentils can be used instead of meat.

## INGREDIENTS
4 large onions
2 tablespoons breadcrumbs
¼ lb/100g minced lamb
handful fresh parsley, chopped
¾ cup/180ml plain yoghurt
2 tablespoons dates, stoned and finely chopped. If using packet dates wash them to remove some of the sweetness.
2 tablespoons walnuts, chopped
1 tablespoon raisins or sultanas
salt and pepper

## METHOD
1 Peel the onions and place them in a large pan of boiling water. Reduce the heat and let them simmer for 15–20 minutes until they are fairly tender. When they are ready take them out and set aside to cool.
2 Now remove the centre section of each onion to leave a shell about ¾ inch/1.5cm thick. In a bowl mix together the meat, pepper, salt, yoghurt, dates, walnuts, raisins or sultanas and breadcrumbs. Fill the onions with this mixture. Keep any that is left over and mix it with the chopped discarded centres of the onions.
3 Now place the onions in an oven-proof dish, spoon any remaining mixture around them and then cook for 1 hour. Garnish with parsley and serve with boiled barley, cracked wheat or rice.

# 5
# Gendered fields

## Introduction

This chapter examines how differential access to resources by gender influences food production and distribution and, consequently, patterns of hunger and malnutrition.

> Governments and international agencies also continue to blithely ignore the mounting evidence that women, as the main providers of basic needs, are crucial to understanding and resolving the crisis of rural reproduction in the Third World.
>
> (Sen and Grown, 1987, 58)

Numerous categories may be employed to analyse malnutrition and food insecurity, from the very general, 'the poor', to the more refined 'landless' or 'refugees'. What is the rationale for devoting a chapter to the category 'gender'? There are three justifications: the relative invisibility and neglect of women as bread-winners in the literature about food security and hunger; their prominence in statistics on malnourishment; and their crucial role in influencing the nutritional status of children. It may be justified therefore with reference to production and consumption factors.

Until Boserup's classic study (1970), the economic role of women in agriculture was ignored; in more recent analyses the significance of their labour in this sector is still frequently minimised. Among the most important specific issues beginning to receive attention are the vital role of female farmers in the

South; the gender division of labour in agricultural production, especially the crucial role of female labour in subsistence and cash crop production; and differential access to agricultural resources and extension services.

Women's labour in rural or urban areas in not of course confined to the agricultural sector, and another often neglected theme is the numerous contributions to livelihoods females make in other economic activities – these are extremely diverse, although they often display geographically specific patterns. In all regions, however, 'much of women's work remains unrecognised and unvalued' (UNDP, 1995, 87). Because female labour in the South is often in the unorganised (informal) sector and in household and community work, it is often ignored, although things are improving. The variety of ways that women contribute to their own and their households' survival is suggested in the discussion below.

The second justification is dealt with in Chapter 2, but to recap, in some of the world's regions where hunger is most serious, women's nutritional status is worse than that of males and is reflected in their higher morbidity and mortality rates. Sen's anouncement that '100 million women are missing' indicates the seriousness of gender bias in these regions. Globally, life expectancy at birth is sixty-five years for females and sixty-two years for males. In some populations there are about 106 females for every 100 males. In sub-Saharan Africa, there are 102 females for every 100 males. Gender discrimination is obvious where this sex ratio is reversed, as in China and South and West Asia, where there are only 94 females for every 100 males. Applying the sub-Saharan ratio of females to males, the shortfall of females is particularly pronounced in China, where some 49 million women appear to be missing. Adding this figure to the shortages in North Africa and South, Southeast and West Asia, as Amartya Sen has done, leads to an estimate of more than 100 million women 'missing' (UNDP, 1995, 35; also Momsen, 1991, 7–13).

These patterns reflect gender status and are linked to patterns of consumption within households: consumption of basic goods such as food but also entitlements to other resources such as health care. This is allied to the last justification. There is evidence from across the South that the status of mothers is closely correlated to the nutritional status of their children – not a breathtaking conclusion perhaps? However, what is important is how slight changes in the mother's circumstances can have important implications for her children's nutritional status and health. Also, child nutrition is positively correlated with the size of the mother's income; this is not the case for the father's. These themes form the basis of this chapter and its structure. Its main premise is that attempts to understand hunger which are blind to how gender influences its incidence are fated to be partial. This chapter begins to indicate how a gender-aware perspective informs analyses of hunger.

## New measures of gender inequality

Since 1990, the United Nations Development Programme has published the human development index (HDI). The HDI is a composite index which measures the average achievement of a country in basic human capacities (it is based on life expectancy, adult literacy rates and school enrolments). It is more helpful than purely economic measures like GNP per capita because it attempts to measure the improvements in people's well-being. The HDI indicates whether people lead a long and healthy life, are educated and enjoy a decent standard of living. The optimum HDI is 1.0, so the closer a country comes to 1 the better the quality of life for the majority of its people. There may be significant differences between rankings of countries by their GNP per capita statistics and their HDI rankings. Table 5.1 makes some interesting comparisons. It is clear that that some governments are better than others at ensuring that economic gains are translated into standard of living improvements; Madagascar and Costa Rica, for example, have been more successful than the United Arab Emirates or Tunisia.

**Table 5.1** HDI and GNP rankings, 1992

| Country | HDI | HDI rank | GDP per capita rank | HDI – GDP rank* |
|---------|-----|----------|---------------------|-----------------|
| UAE | 0.861 | 45 | 4 | −41 |
| Madagascar | 0.432 | 135 | 165 | 30 |
| Costa Rica | 0.884 | 28 | 60 | 32 |
| Tunisia | 0.763 | 75 | 66 | −9 |
| Sri Lanka | 0.704 | 97 | 102 | 5 |
| Philippines | 0.677 | 100 | 108 | 8 |
| Indonesia | 0.637 | 104 | 99 | −5 |
| China | 0.594 | 111 | 123 | 12 |
| India | 0.439 | 134 | 141 | 7 |

*Source*: United Nations Development Programme (1995, 18)
*Note*: * A positive figure shows that the HDI rank is better than the (real) GDP per capita rank, a negative the opposite.

Although an improvement on purely economic measures, like GNP per capita, the HDI also suffers from being an aggregate figure which indicates the average condition of all people in a country: it does not capture distributional inequalities by gender, class, ethnicity, age and caste, for example, which will be very marked in most cases.

The launch of the United Nations Decade of Women (1975–85) helped to raise academic and political awareness of the role of gender in all areas of development policy and practice. Even conservative bodies like the World Bank and the International Monetary Fund now recognise that gender has theoretical and practical implications for development. In the last twenty years, considerable effort has been devoted to improving our understanding of the significance of gender, and increasingly data are compiled that facilitate gender-aware analyses.

The gender-related development index (GDI) measures achievement in the same basic capabilities as the HDI but is adjusted to reflect disparities between women and men. The greater the gender inequality, the lower the GDI compared with the HDI. The GDI is the HDI adjusted downwards to reflect disparities between sexes. Because gender inequality exists in every country, the GDI is always lower than the HDI. Table 5.2 compares HDI and GDI indices for selected countries.

**Table 5.2** HDI and GDI values, selected countries, 1992

| Country | HDI | GDI |
| --- | --- | --- |
| Sweden | 0.929 | 0.919 |
| Barbados | 0.900 | 0.878 |
| Thailand | 0.828 | 0.798 |
| Mali | 0.222 | 0.195 |
| Sierra Leone | 0.221 | 0.195 |
| China | 0.594 | 0.578 |
| Saudi Arabia | 0.762 | 0.514 |
| Spain | 0.930 | 0.795 |
| Hong Kong | 0.905 | 0.854 |
| United Kingdom | 0.916 | 0.862 |

*Source*: United Nations Development Programme (1995, 76–7)

It is important to recognise that GDI values are not correlated to levels of affluence. They may be relatively high in the North and relatively low in some countries of the South, or vice versa. This reflects the fact that gender equality characteristics for any country are not correlated to income levels; low-income countries may have relatively high degrees of gender equality, and similarly high-income countries may have high degrees of gender inequality. In countries where the HDIs are very low, and where the GDI is even lower, the situation for women is most bleak; Mali and Sierra Leone fall into this category.

## Women as economic actors

Research in the last twenty years on gender and development may be flawed in many respects but it has certainly helped to make women visible, not least

**Plate 5.1** Cultivating a *dambo* in Zimbabwe

**Plate 5.2** Maize farmers in Zimbabwe

**Plate 5.3** Women harvesting olives in Tunisia

**Plates 5.1–5.4** Female farmers and labourers

Female farmers are important producers of subsistence and cash crops in the South. Frequently, their productivity is constrained because their access to capital, credit, land, labour and decision-making in general is limited. For example, access to *dambo* areas in Zimbabwe is frequently determined by male community heads and is very politicised, as indeed are the decisions about what is planted and who supplies the labour. Entitlements to different rural assets are governed by a hybrid of traditional and more recent tenurial rights and obligations; female entitlements are seldom secure or equal to those of males. Female labour, young or old, family or waged, is often vital to income generation (Plates 5.3 and 5.4).

*Photos*: Plates 5.1 and 5.2, Dr Jennifer Elliott, University of Brighton; Plates 5.3 and 5.4, Dr Hamish Main, Staffordshire University.

**Plate 5.4** Woman winnowing guinea corn in Kano state, Nigeria

in the statistics upon which policy is based. In contrast to prevailing perceptions there are millions of female peasant farmers in the South. Momsen considers the diversity of women and work in rural areas in the South and concludes, 'In general, farms run by women tend to have poorer soil and to be smaller and more isolated than those cultivated by men' (Momsen, 1991, 50). While a female may often be the sole cultivator of a plot, in most countries of the South women are not legally entitled to own land and frequently they have limited rights to use the land. In terms of entitlements, therefore, females are very often literally not entitled to own the land that they manage. This has numerous negative implications, from the inability to use the land as collateral to their being easily evicted by its legal owner (Box 5.1).

**Box 5.1**

---

### Village women pack up and leave

In Kenya, more and more women who have lost their land are migrating to the towns. Take Maria Mathai, for instance. Three weeks after her husband died, twenty-eight-year-old Mathai found that she and her four daughters had no rights over the five-acre farm the family had worked at Chaka, on the slopes of Mount Kenya. When the title deed arrived at the end of last year, it was in the name of her husband. A widow with no son, she was not entitled to inherit the farm, despite the fact that she had worked it more than her husband. The chief of her husband's clan has asked her to move out of the village. She is lucky, however, because her plight caught the attention of the national media and funds are being raised to help her to buy a new farm. Other women in similar circumstances must stay and work as casual labourers or set off with their children to the cities.

In an effort to stem the tide of rural–urban migration the Kenyan government is working on land reform legislation that converts customary land tenure into freehold farms. The idea is that if farmers have security of tenure production will increase, because farmers would have incentives to undertake permanent improvements. But many feel the new legislation is not helping women.

'The legislation in actual fact suggests that access to land for women as a group is being systematically eroded,' says Elizabeth Nzioki, who is studying the effects of the land reform on women's access to and control of land for food production. Nzioki points out that land titles are

---

Box 5.1 (*continued*)

being transferred almost exclusively to men, 'thus transforming women's traditional access to land to that of simple labourers.' And yet, she says, 'most of this land has been inherited, not bought, by the men. In this country, women do not inherit land but all along, we have been depending on our husbands or in-laws to give us land to farm.'

The title deed is an important asset in Kenya, especially to get loans. Most women cannot get loans because they have not got title deeds. The land tenure reforms have also led to commercialisation of agriculture. But even as income goes up with the growing of cash crops, the burdens on women increase. They are now expected to work for longer on cash crop production, which means they have less energy and time for food crop production. Profits from their labour in cash crop production, however, go to the men, who have the title deeds, and the women then have little influence on how this money is spent. The title deed, in short, has become a legal instrument empowering men to gain almost complete control over land and over women's labour.

*Source*: The Women's Feature Service (1993, 61–2).

In sub-Saharan Africa much of the rural labour is female. Recent results from a survey in Kenya found a number of contrasts by gender and age. Women in rural Kenya work on average fifty-six hours a week, while men work an average of forty-two. These gendered patterns are found in earlier age groups too; if time for education is counted, girls between eight and sixteen years work forty-one hours per week and boys of the same age thirty-five hours. Women shoulder the heaviest burden in housework, firewood and water collection (ten times the hours of men) and again in the younger age groups, girls work on household duties 3.7 times more than boys. In households that farm cash crops, women work the longest hours – sixty-two hours a week – and as this sector expands these hours will expand.

Despite these realities, access and ownership of the land is controlled by men. As European colonialism proceeded in the late nineteenth and early twentieth centuries, capitalist enterprises appeared, commonly mines and plantations, and males were forced to migrate from rural areas to waged work in these sectors. Colonial tax policy worked in conjunction with the demand for labour to produce skewed sex ratios in both urban and rural areas: a female bias in rural areas and male biases in industrial and urban areas. Colonial and post-colonial agricultural policies have ignored this sector and there must be truth in Taylor's assertion that 'The acute food shortage in Africa, while dramatically and tragically exacerbated by drought and war, may be due in large part to the way women have been systematically excluded from access to land and from control of modern agriculture in that region' quoted in Tinker, 1990). Access to land is an important source of entitlements in the South and securing legal possession of land for women is crucial to improving food security. Generalisations about sub-Saharan Africa are problematic because of its diversity but several common themes emerge from the glimpse of Joyce Kayaya's life as a Zambian farmer (Box 5.2).

Box 5.2

### Joyce Kayaya, a Zambian farmer

Fully 85 per cent of Zambia's farmers work, like Joyce, on a very small scale. Yet they contribute over 60 per cent of the country's maize production. She farms twenty hectares but, in common with most small farmers in Zambia, she does not hold the title deeds to her land. It is ancestral land held by a local chief. At her *nyumba* (homestead) she recalls that she used to grow just enough maize to feed her family but in 1984 she took out a loan from the credit union. She began to grow and market more maize and expanded again in 1987, when she became a contract farmer under Global 2000 'to help improve my farming methods. Others came to see me and to learn.' She has diversified to make better use of the rains, growing cabbage and other vegetables as well as maize.

Joyce was puzzled by the 1986 food riots in Zambia. 'The food riots we heard about in the towns were simply a result of laziness, laziness in the sense that if there are so many other people like me who produce so much maize, why is there such a shortage that the government has to buy from other countries?' She points to the fact that each year there are no funds to transport the maize to the depots and if it gets there there are poor storage facilities so that it rots when the rains come. She also thinks that there is enough land in Zambia to grow food and that people should not migrate to the cities unless they have a job there and can provide for themselves. Otherwise, she believes, people should stay in the countryside and grow what they need to eat, as she has done since 1965.

Box 5.2 (*continued*)

In 1965 Zambia gained its independence from the British and her husband lost his sight.

At that time we had five children. The oldest was only seven and the youngest a baby. There was nothing for me to do except take full responsibility for cultivating our land.

Her husband managed to make bricks to earn a little extra funds for the household, and in the end

we raised enough money to buy two oxen to help with the ploughing and to pay for uniforms and schoolbooks for the children. In those days I did all the work myself. I hitched the plough to the oxen and ploughed with a child on my back.

Joyce also did all the cooking. The family diet consists mainly of *nshima*, a hard maize meal porridge, eaten with rainfed vegetables like rape, cabbage, okra, cassava leaves and mushrooms. Sometimes they buy beef or goat meat, or slaughter a chicken; they reserve their cattle for weddings and funerals. Joyce sometimes makes a drink called *chibwantu* from crushed maize and the roots of a common tree.

Joyce has nine surviving children; all, except one, are independent now and visit her occasionally and bring presents. One of her sons, Obed, stayed on the farm. Joyce's house is on a hill and overlooks their large maize field. They have a small orchard of pear, peach, mango and mulberry trees, which provide enough fruit for the family. They hope one day to build a pump and grow vegetables on a commercial basis. 'I really don't have any complaints,' says Joyce. 'I love my smallholding and thank God for what he has given me.' Joyce is an inspiration to farmers in her area. It is unfortunate that their efforts are not rewarded by more efficient distribution of maize.

Low copper prices, high oil prices, debt repayments and the cost of caring for refugees from neighbouring countries have all burdened Zambia's economy, but many of its people are more likely to pin the blame on bureaucratic inefficiency. Wherever the responsibility lies, unless something changes the country will remain a maize importer.

*Source*: Namakando (1991, 10–12).

Women may be peasant farmers but they are also often farm labourers. Statistics about female labour in general are unreliable and for various reasons may be most unreliable in the agricultural sector (Momsen, 1991, 47). Gradually, however, research is revealing the significance of female agricultural labour in all world regions. Even in cultural contexts where women are theoretically 'confined' to the household, they are often, in practice, working in the fields, in Bangladesh, for example. Mies and Shiva (1993, 234) estimate that in India, agriculture employs 70 per cent of the working population, and about 84 per cent of all economically active women. Contrasts within South Asia are vast but the following figures confirm the role of female labour in family production: in the Himalaya, a pair of bullocks works for 1,064 hours, a man for 1,212 and a woman for 3,485 hours a year on a one-hectare farm; in Kerala, women have a predominant role in all processes except ploughing, which is exclusively men's work (more than two-thirds of the labour input is female).

The gendered division of labour in agriculture helps to explain the mixed impact of new technologies. The introduction of new technologies is always associated with differential impacts; some social groups benefit disproportionately and others may suffer disproportionately. Certainly, there will be a shift in social relations and generally speaking the more significant the technology the greater the differential impacts, by class and gender. Conventional agricultural extension services have ignored female producers; prejudice is still very high for a variety of reasons and explains the failure of many agricultural development programmes. Often technologies that are promoted are inappropriate or actually detrimental to agricultural production or local nutritional status.

## Female entitlements and the unorganised (informal) economy

It is important to stress that in rural and urban areas female entitlements are based on a multitude of activities, the bulk of which are in the unorganised (often referred to as the informal) sector. Although conventional economic analyses tend to regard this sector as marginal it is in fact the most crucial in terms of entitlement provision for all the world's poor. It is expanding and increasingly significant for the poor's survival. This variety must be appreciated for effective policy interventions, but activities vary by gender, age, season and region. The vast majority of women work in the informal sector, where they may:

> market smoked fish, raise chickens, grow food crops, extract oils, keep bees, make bricks, process salt, brew beer, sew school uniforms, make baskets or pots. They learn to make carpets, roll cigarettes, manage inns and restaurants, mine stones and gems, and do carpentry work. They are domestic servants

and they carry headloads. Some sell drugs and others turn to prostitution. Those endeavors and countless other small enterprises are estimated to employ up to 70 per cent of the labour force in the rural and urban areas of developing countries.

(Synder, 1995)

Activities in the unorganised sector are typically labour intensive, use simple technology, require little capital, are insecure and are based on local markets. Women's income from these activities is not a supplement to household income; rather it is a vital component of household budgets and vital to their members' survival. In the past and present these activities are often considered marginal by policy-makers. Workers in this sector suffer from lower wages and greater insecurity than those employed in the formal sector, so any attempt to reduce poverty and increase food security must appreciate the role of these activities to survival strategies. Limited but significant success is evident from some projects which target female activities in this sector.

## Female remuneration in the organised (formal) economy

Whether in the formal or informal sector, women earn less than men. Various explanations are offered for this state of affairs. Fundamentalist religious doctrine – Christian, Judaic or Islamic – emphasises marriage and motherhood as the ideal activities for women; their role as economic actors is consequently often viewed as marginal, as supplementary and therefore not worth a high income. Even in more secular contexts, kinship, state structures and employers view motherhood as the most important way for women to earn respect and status; the image of 'the mother' is revered. Images of working women are much less attractive and in many cases are negative: economically independent women are not revered; they are sometimes reviled, especially where their independence is viewed as a threat to prevailing patterns of patriarchy. Prevailing ideologies about females attitudes and aptitudes are also used to justify low female wages and male bias in high-prestige jobs. Given the already important role of female wages to household entitlements and the growth of female-headed households, the relevance of female remuneration to entitlements and food security is obvious.

## Intra-household entitlements

The control and allocation of resources within the household is a complex process which has to be seen in relation to a web of rights and obligations.

**Plate 5.5** Women harvesting mussels from wetlands outside Calcutta

**Plates 5.5–5.8**: Female employment and the informal economy

Throughout the South the great majority of women contribute to what is called the informal or unorganised sector. Although more important for poor people's entitlements than the formal sector, it continues to receive scant policy attention. Plate 5.5 shows a woman gathering mussels, which will be marketed in Calcutta. She is working for piece rates and the bulk of the profits will go to the few large landowners who own the wetlands area. As you can see she has a child with her. Throughout the South working women very often have their children with them as they work.

Female street traders are a common sight in Latin America and are an important link in the urban food distribution system, selling vegetables, herbs and snacks; sometimes, as in Plate 5.6, they sell crafts they have made to tourists. Again, they are often accompanied by their children.

**Plate 5.6** Women selling craft work to tourists, small town, Ecuadorian Andes

**Plate 5.7** Women selling *kosai* (chick pea snacks) in northern Nigeria

**Plate 5.8** Female vegetable trader, Jos Plateau, Nigeria

*Photos*: Plate 5.5, Dr Jennifer Elliott, University of Brighton; Plate 5.6, Katy Hawkins; Plates 5.7 and 5.8, Dr Hamish Main, Staffordshire University.

> The management of labour, income and resources is something which is crucially bound up with household organisation and the sexual division of labour.
>
> (Moore, quoted in Kabeer, 1994, 95)

Our analysis into the processes which influence access to entitlements, and therefore food, cannot stop at the door of the household. Inside households, ideas and practices influence how available resources are distributed. There are marked intra-household welfare inequalities because power relations exist within households too. Employing gender-aware analyses helps to explain many of the observed discrepancies in household well-being and these are considered next. Two assumptions are selected for examination: that household heads are male and that households are harmonious oases protected from the conflicts that characterise virtually all other social institutions.

One long-lived assumption is that 'households' are always headed by a male, a father or husband, who is the prime bread-winner. In fact, a large and, evidence suggests, increasing number of households survive on female-earned income. Females in these households may be *de facto* or *de jure* heads but in all cases they are the primary income-earners.

The composition and circumstances of these households are very diverse as, indeed, are the factors that explain their occurrence. They may be headed by widows, divorced women, women who are deserted or who chose to desert, women who chose to live without a permanent male partner, female migrants or refugees, etc. The number of these households in many regions of the world is increasing as socio-economic and cultural conditions change.

Most social and economic policies are designed on the assumption that males are the heads of households and the principal bread-winners, which has serious implications for household entitlements. The gender of the person owning resources affects how resources are allocated and consequently the short- and long-term survival strategies of the household. Research from various geographical regions suggests that income controlled by female household members is more likely to have positive effects on household nutritional status. Increases in female income are more likely to be correlated with improvements in the diets and health of household members than are increases in the income of male household members. This introduces the second assumption, that households are oases of calm, where conflict is rare.

In some societies the household is the only realm where women are allowed some autonomy and authority and it may serve as a relatively safe refuge. However, within many households there is conflict over how resources are used. Conflicts occur over the use of household labour and the use of income earned by various household members. Most of us experience daily, or remember, conflict over what various family members are expected to do, their responsibilities, and those things we presumed were legitimately ours, our rights. How these conflicts are resolved reflects intra-household patterns of power; although negotiation occurs and bargains are struck, the will of the most powerful household members generally prevails and their status is maintained. Households in the South struggle over much the same issues, except in circumstances where these decisions may have greater significance: who enjoys a better diet; who helps with housework and who does homework; who helps in the garden plot or collects fuelwood; and who travels to school or works for wages (Box 5.3). In these contexts decisions about rights and responsibilities within households become crucial and in most circumstances reflect differential power relations between males and females. There is a marked difference in the proportion of earnings that women and men spend on personal consumption; gender ideologies commonly expect women to behave altruistically and 'allow' men to be less so.

**Box 5.3**

---

### Contrasting cases: gender relations and household expenditure

#### An Egyptian example

This information is drawn from a study undertaken in a low-income neighbourhood in Cairo in 1983–84. Nuclear families were the norm in this community; men usually worked in the modern sector of the economy and how their income was spent was very contentious. The survey concluded that the more open men were about their earnings the less strife there was in the household and that women were more prepared to pay for their children's education and private lessons. The most visible difference between families in which women had an independent income and those in which they did not was in how women and children were clothed. This quotation indicates the contrasts in diets and mobility among household members:

> He eats chicken and kebabs for lunch and dinner; we do not see meat more than once a week. He spends the rest of his time in cafes and cinemas, while we cannot leave the neighbourhood from year to year.

---

Box 5.3 (*continued*)

In most families, men were better dressed than their wives, especially those who wore Western dress. There was a great deal of resentment among the women about the amount of money men spent on themselves. Generally there were fewer arguments when women had access to an independent income, but arguments were regarded by most women as pointless because their husbands would simply leave the house, spend time with friends, and return only to go to bed.

We must understand this example with reference to the traditional doctrines of sexual segregration. Only men were allowed to 'work' (earn money), especially in the modern sector. Opportunities for women without a husband or an independent income were very bleak. In this case, men in low-income jobs had adopted Western clothes but adhered to traditional gender roles. Contrasts are available within Egypt from different classes and, as elsewhere, gender and class are important determinants of circumstances in the community and the well-being of its members.

### An Indonesian example

A study of lower-middle and middle-income married women in Jakarta contrasts with the case above. Two external factors influenced women's and men's behaviour and strategies in relation to household finances. In Indonesia, women's participation in the workforce is high, about 30 per cent, and divorce rates are also high, also about 30 per cent. These circumstances help to make negotiations about household expenditures more equitable. The fact that women are wage earners benefits their families, but it is also of long-term benefit to them personally. They develop the skills and resources to support themselves and family members in the event of a divorce or widowhood. Undoubtedly in Indonesia, as in the West, when women enjoy more economic opportunities their options increase and they are less willing to maintain relationships with partners who are selfish. Survival, of themselves or their children, ceases to depend on being part of a traditional male-headed household.

*Source*: O'Connell (1994, 57–8).

An extreme manifestation of divergent household entitlements is where girl children eat last and least. In regions of the world where female infant mortality is higher than that of males (South Asia, East Asia, the Middle East and parts of Africa), female infant malnutrition rates are higher too, suggesting that less care is given to female babies and young girls. Evidence is now available which confirms differential feeding and care patterns between boys and girls. The perception is widespread that infant boys are fed more adequately than infant girls in poor areas, suggesting a gender bias in favour of male children. The regional patterns described below suggest some patterns but must be used cautiously.

In sub-Saharan Africa, the prevalence of moderate and severe under weight (weight for age) is 17 per cent for girls and 32 per cent for boys, showing a better situation for girls. The reverse is true in Latin America and the Caribbean: 31 per cent of girl children are underweight, compared with 17 per cent of boys. In Bangladesh, malnutrition is experienced somewhat more by girls than by boys: 59 per cent of girls and 56 per cent of boys suffered chronic malnutrition, and 10 per cent of girls and 7 per cent of boys suffered from acute malnutrition. Evidence of nutritional deprivation among women and girls appears most starkly in their reproductive years: 77 per cent of pregnant women from middle-income households and more than 95 per cent of those from low-income households weighed less than the standard 50 kilograms. In India's rural Punjab, poverty takes a bigger toll on the nutrition of girls than that of boys: 21 per cent of the girls in low-income families suffer from malnutrition, compared with 3 per cent of boys in the same families. In fact, low-income boys fare better than upper-income girls (UNDP, 1995, 35). The blatant misuse of amniocentesis in India and China, for example, is an extreme manifestation of this gender bias; in this instance female foetuses face discrimination.

In the context of poverty, investments in male children give better returns. To recognise this fact is not to condone it, rather it is to suggest the deeply embedded nature of gender bias: institutional structures at all levels of society conspire to prejudice a girl's prospects *vis-à-vis* a boy's. As emphasised previously, contrasts within countries may be marked: a girl child's chances of survival may differ markedly between regions, or within regions by class, caste, or ethnic or religious group. It is important to stress this, because these variables may be most significant. Education and socialisation processes tend to reproduce themselves and there are many of both genders who have interests in maintaining the status quo, but there are signs that some of the extreme forms of prejudice are being challenged.

## Conclusion

The discussion in this chapter describes the problems women face in command-ing resources, which results in the higher incidence of malnutrition among girls and women than among boys and men. It would be a serious error to finish before describing the successful efforts females are making to challenge discrimination and disadvantage at all levels of society, from the household to globally. It is important to note that the situation is improving; Table 5.3 shows the changes in GDI values and indicates that progress has been achieved.

**Table 5.3** Changes in average GDI values, 1970–92

| Group | GDI, 1970 | GDI, 1992 | % change |
|---|---|---|---|
| All countries | 0.432 | 0.638 | 48 |
| Industrial countries | 0.689 | 0.869 | 26 |
| Developing countries | 0.345 | 0.560 | 62 |

Efforts by women to improve their circumstances have always existed but have increased and become more voluble in recent decades. Strategies have varied since the initial emergence of gender and development debates in the 1960s, but in the 1990s it is apparent that women's formal and informal lobbies, organisations and movements have evolved in diverse regions and with a range of ambitions and objectives. These are the most important legacy of the debates, policies and practices initiated in the 1960s. Whatever the initial aim of these women's groups they have provided space for women to achieve a degree of independence and political expertise that may result in their empowerment (see Chapter 8 for fuller discussion). This term is contentious. Here I mean simply that through participation in these organisations millions of women have improved their self-esteem, gained some political experience, and appreciated that they can make a difference to their lives.

Empowerment is often associated with grassroots organisations (GROs) and non-governmental organisations (NGOs), both of which have proliferated in the 1980s and 1990s. GROs usually emerge with a specific objective, but many develop momentum, which ensures their growth and survival when the initial battle is concluded. Women's groups in the South have emerged to fight for reproductive rights, environmental issues, employment-related causes, mother and child health programmes, nutritional improvements, improvements in sanitation and water provision, and to expose the extent of domestic and state violence, rape, and multinational exploitation, to mention a few. It has been difficult deciding which examples to include here: I have opted for three examples with obvious salience for improving female entitlements (Box 5.4).

Box 5.4

---

**Female empowerment**

## The Grameen Bank

The Grameen Credit Union in Bangladesh started out as a poverty eradication project in 1976 and was based on research that had established that rural people survived by all sorts of self-employed activities, not just by earning wages. Because the rural poor had no way of getting credit, their efforts to increase incomes from these activities were limited. The Grameen Credit Union provided credit for the poor and helped them to increase their earnings from self-employed activities. It became an independent national bank in 1983.

Although it started by lending primarily to men, and the majority of its staff are male, its main clients (90 per cent) are female. The Grameen Bank is improving opportunities for poor women to increase their incomes and thereby their self-esteem and independence in a country where gender differences are very marked. Some consider it limited in its scope and empowerment potential, but fundamentalist religious opinion in Bangladesh is very opposed to the bank because it is viewed as a dangerous threat to traditional family and societal norms.

## The Self-Employed Women's Association of India, SEWA

This association emerged to address the needs of female workers in the unorganised (informal) economy in India. Although the trade union movement had a women's wing as early as 1954, it targeted women only as dependants of male mill workers, not as independent economic actors. A survey in the early 1970s uncovered the existence of large numbers of women in the textile economy who, as women tailors in the unorganised sector, were ignored by the organised unions. The SEWA was formed to represent the interests of women who were home-based workers (outworkers), petty traders and casual wage labourers. Statistics are problematic but perhaps as much as 90 per cent of employment in India is in the unorganised sector, also known as 'informal', 'unprotected', 'unregistered', 'marginal' and 'black economy'.

Originally an urban-based organisation that focused on narrowly defined work-related issues of women in textile production (legal aid,

Box 5.4 (*continued*)

credit provision, workers' rights, better working conditions and higher wages), SEWA has expanded its focus to encompass work with poor rural women and a much broader range of issues, which include health and child-care cooperatives, life-insurance schemes, and maternity benefit schemes. Recognising the obstacles poor women have in using formal banking systems, SEWA has also established its own bank, staffed by people sensitive to the special constraints of poor women. Its emphasis upon participatory approaches has meant that many issues, previously shrouded by the 'culture of silence', have been exposed and are receiving action, such as domestic violence, sanitation issues, etc. SEWA is an effective organisation articulating the concerns of poor urban and rural women. Staffed by them it has become an important organisation struggling to improve female entitlements in their homes and work places.

## Indian Widows Conference

In April 1994, a conference in India highlighted the plight of the approximately 33 million widows in India. Widows in India are often denied their rights; for example, although they are legally entitled to inherit at least part of their deceased husband's property, only about half exercise that right because they fear violence and retribution from other family members if they assert their claims. One of the most important aims of the conference was to help foster the self-confidence of the widows who attended so that they might effectively challenge some traditional norms. The following report in *The Hindu Magazine* suggests the changes that the conference initiated:

Throughout the week they [widows] came to realise many things about themselves and their lives – especially how much they had internalised society's perception of them as daughters, wives, mothers and widows (their identity invariably defined in terms of their relationship to men). The workshop aimed to change their self-perception as objects of pity, unfortunate women who had lost their husbands and now had to beg for help from their families or sops from the state. They were encouraged to see themselves as persons who had a right to exist even if their husbands were dead, and as citizens who had a right to resources – such as land, housing, employment, credit and ration cards

Box 5.4 (*continued*)

> – which would enable them to live and bring up their children (if any) with dignity and self-respect.
>
> That the process they went through in that week was a transforming one was evident on the last day, when they got together one more time in a symbolic and moving ceremony which reinforced their newly acquired sense of unity and strength. Before they bid each other farewell, they shared their individual decisions about what they would do to carry the message of the workshop back to their villages. While their promises to each other – most of them related to wearing forbidden things like bindi, sindoor, bangles or coloured clothes – may seem trivial to many of us, they represent huge strides in their march towards self-confidence and an identity that is all their own.
>
> *Source*: Nussbaum and Glover (1995, 14).

## Key ideas

1 Entitlements vary by gender and influence the incidence and patterns of hunger. Women's nutritional status in many parts of the South is worse than men's.
2 Gender inequality exists in all societies and is not necessarily correlated with a country's wealth. It is reflected in some statistics – sex ratios, morbidity rates, earnings and literacy levels – but its persistence is due to factors that are more difficult to quantify, i.e., ideologies about the nature of femininity and society.
3 Women in the South are important economic actors in the formal and informal economic sectors; their earnings are vital to household entitlements. Female entitlements may be the only source of household entitlements in many cases; the 'bread-winner' is often a female.
4 Improvements in a mother's entitlements are often directly reflected in improvements in children's nutritional status; this is not always the case when men's entitlements improve.
5 Decisions taken by a household about the use and distribution of resources are very often contentious; all household members are not given equal rights.
6 In many countries women have relatively few legal, civil or political rights compared with men, which means their entitlements are very restricted.

# India

## MIXED VEGETABLE CURRY                                   Serves 4

Although India is among the world's top ten industrial countries, millions of Indians cannot afford enough to eat. Yet the largely Hindu population, as vegetarians, are among the most efficient food users in the world, eating vegetables and pulses with chapatis or rice – a more direct means of gaining nutrients than by feeding grains and pulses to animals first, as we tend to do in the West. This curry is a good way to use up vegetables from a previous meal, or it can be made from scratch using potatoes, carrots, zucchini/courgettes, peas, okra – whatever you like.

## INGREDIENTS
4 tablespoons/50g margarine or ghee
1 onion, sliced
1 tablespoon curry powder
1–2 cloves garlic, crushed
2¼ cups/350g vegetables, diced and parboiled
1 green bell pepper, chopped
1 tablespoon desiccated coconut
a little water or stock
½ teaspoon turmeric
salt and pepper

## METHOD
1 First of all, melt the margarine or ghee in a large pan and cook the onion and garlic until they are lightly browned. Sprinkle in the curry powder and turmeric; cook for 2 minutes, stirring frequently.
2 Put in the bell pepper and let it cook for a few minutes before adding the other vegetables, coconut and enough water or stock to cover the base of the pan and prevent sticking.
3 Now cover and simmer until the vegetables are tender, adding more water or stock if the mixture looks too dry. Season with pepper and salt and serve with rice, chutney and yoghurt.

# 6
# Sub-national perspectives

## Introduction

Understanding global hunger requires an appreciation of the global and national contexts within which millions of people are denied access to food. The probability of suffering from hunger is greater in Southern countries than in Northern ones but contrasts between states in the South are enormous too; analysis at the sub-national level reveals more contrasts. Not all regions, or groups of people, have the same probability of suffering hunger within developing countries, that is entitlements display marked intra-national contrasts which are often crucial to explanations of the incidence of chronic or acute hunger. Some important factors operative at the sub-national level have been examined already: Chapter 5 explored some of the most important connections between gender and hunger, and contrasts between urban and rural populations and their command over food have been considered already. This chapter examines other variables, operative at the sub-national level, which influence command over food, namely regional contrasts, minority status and contrasts between households.

## Regions

Some regions have markedly better food security than others. Regions which are economically or politically marginal have a higher incidence of hunger than more prosperous regions. Some regions become marginal because of the nature of their integration into the national economy, their environmental endowments, and their past and/or current economic and political relations to more powerful regions. I open by considering a region which is distinct because of its success in reducing hunger.

Regional contrasts within India are very marked and reflect federal policies and contrasts in the entitlements and commitment of local states. India has eliminated famine since independence and has become self-sufficient in food production. These achievements reflect the federal government's commitment to food security shown in its various interventions in the economy, including the establishment of the national Public Distribution System. Famine disappeared despite rapid population growth. Unfortunately, available evidence suggests that the government has been less successful at reducing the incidence of chronic hunger, although some states have been much more successful than others.

Most of India's 1,022 million people are still rural (74 per cent in 1992) and the bulk of the poor are rural people, but as rural–urban migration increases the percentage of the population classified as urban is growing and so too is extreme urban poverty. The statistics in Table 6.1 indicate that access to basic services is better for urban populations; in that these will be important indicators of general standards of living, we can cautiously assume that food security would reflect the same pattern, i.e., be better in the urban areas.

**Table 6.1** India, 1992: urban–rural discrepancies in access to basic services

| Service | % of population | |
| --- | --- | --- |
| | *Urban* | *Rural* |
| Health | 100 | 80 |
| Safe water | 85 | 78 |
| Sanitation | 62 | 12 |

*Source*: United Nations Development Programme (1995)

Despite some optimistic, and contentious, official statistics about economic expansion and poverty reduction, all observers agree that poverty in both rural and urban areas is serious and that the absolute number of very poor people is growing in both. Economic expansion in some sectors and its fruits, enjoyed by some classes, obscures the fact that a heterogeneous population of poor people may have been suffering a relative decline in their material circumstances. In rural areas, tribal populations, landless people, small farmers and populations very reliant on common property resources continue to lose entitlements and their access to a decent diet. In urban areas, recent migrants and workers in the unorganised sector who live in shanty towns are most vulnerable to malnutrition. In analysing the regional disparities it is important

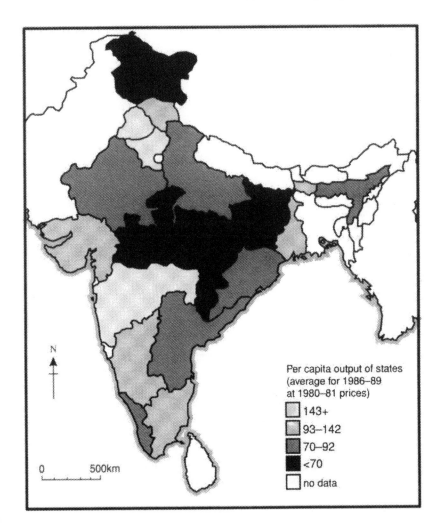

**Figure 6.1** India: regional diversity
*Source*: Bernstein, Crow and Johnson (1992, 145)

**Table 6.2** Regional diversity in India

| State | % of all-India average |
|---|---|
| *Northern Region* | |
| Punjab | 174 |
| Haryana | 143 |
| Himachal Pradesh | 97 |
| Jammu and Kashmir | 35 |
| *Western Region* | |
| Maharashtra | 143 |
| Gujarat | 112 |
| *Central Region* | |
| Uttar Pradesh | 75 |
| Rajasthan | 71 |
| Madhya Pradesh | 32 |
| *Eastern Region* | |
| West Bengal | 94 |
| Assam | 78 |
| Orissa | 70 |
| Bihar | 53 |
| *Southern Region* | |
| Karnataka | 99 |
| Tamil Nadu | 98 |
| Andhra Pradesh | 79 |
| Kerala | 72 |

*Source*: Bernstein, Crow and Johnson (1992)

to note that intra-regional difference may be more significant in some cases. In all regions, other variables which explain the incidence of malnutrition may be more salient, for example, gender, caste or ethnicity.

Analysis of regional per capita output figures between 1986 and 1989 reveals those states which are at each end of the wealth spectrum. The most prosperous states are Punjab, Haryana and Maharashtra, while the poorest are Madhya Pradesh, Jammu and Kashmir, and Bihar. Poverty exists in the wealthiest states, but its incidence is higher in the three poorest states. These patterns are reflected in mortality and morbidity rates, which are higher in the central and eastern regions than in the north or south.

Kerala, for example, has a good record for addressing poverty and its symptoms, including malnutrition. Socio-economic indicators for Kerala are above average for India and are favourable compared with upper middle-income group countries (Brazil, Mexico, Argentina and Venezuela, for example). The reason for this success lies in historical circumstances and the politics of the local state since independence (Sen, 1992). Historically, Kerala has enjoyed a more egalitarian social structure and, even in colonial times, local taxes and rents were less onerous than in many other regions. The major explanatory variable cited by Sen 'is a process of democratization [that] had altered the meaning of the public sphere, with the result that large numbers of people could participate both in the definition of social needs, and in assuring accountability for their fulfilment' (*ibid.*, 276). In this case, a politically organised and articulate electorate in a democracy forced the state to address the problem of hunger.

Lipton and Maxwell (1992) describe how a large and growing proportion of poor people in the South rely on 'marginal farming' for their livelihoods. They estimate that 30–40 per cent of the poor in the developing world are in regions where the land is unreliably watered, ill-drained, hilly or otherwise fragile and eroded. If we add the numbers of urban poor living and trying to survive in degraded or environmentally hazardous areas the enormous size of the problem is clear (Elliott, 1994, 21). With few or worsening entitlements, people surviving in these environments are forced to further degrade their environments, thereby threatening sustainability to secure current incomes. Access to common property resources (CPRs, typically community grazing, gathering, fuelwood rights, fishing rights, etc.) is usually vital to the survival of these populations, yet entitlement to these resources is dwindling. Access is lost because once common lands are being incorporated into private holdings. The same is true for access to fishing grounds. Previously rich, diverse coastal habitats are being depleted by industrial fishing concerns and/or increasingly devoted to marine farming. Produce from these commercial activities does not sustain local populations; rather it sustains distant populations and erodes local entitlements.

This is an appropriate time to introduce the role of climate in food crises. This text rejects the idea that 'climate explains famine or food insecurity' and emphasises instead the role of historically evolved socio-economic and political patterns of entitlement provision. However, granted the primacy of the social character of food security, it is important to appreciate the way that climatic variables may increase vulnerability to food insecurity in some regions and how they may be the proximate cause of famine. 'Seasonality continues to undermine household food security, and drought regularly triggers food crises, particularly in sub-Saharan Africa. Climate cannot therefore be totally ignored' (Devereux, 1993, 44). Some of the connections between seasonality and drought and food insecurity are now reviewed.

The relevance of seasonality is most marked in semi-arid areas, where access to clean water varies markedly during the year; the timing as well as the amount of rainfall is crucial in these regions. The issues raised here are pertinent to all regions, but in semi-arid regions they are more acute and have direct implications for food security. The implications are several: in these regions

**Plate 6.1** A farmer in Zimbabwe battles against soil erosion

**Plate 6.2** Soil erosion in a semi-arid region in Tunisia

In Plate 6.2 *bunds* (earth banks) are visible in the gully; these are built to try to reduce soil erosion.

*Photos*: Plate 6.1, Dr Jennifer Elliott, University of Brighton; Plate 6.2, the author.

the cost of water is often high and volatile, so people just managing to survive may be plunged into extreme poverty. As the cost of this basic need increases, they are forced to devote a higher percentage of their income or energies to providing water for themselves or their livestock. If rains are late, or at the other extreme if floods contaminate clean water sources, the poor are forced to pay inflated prices for clean water. To exacerbate matters the cost of water often increases at the same time as food prices rise, so reducing the poor's ability to access both basic needs. Where water is at a premium people reduce their use of it and sanitation and hygiene tend to suffer, so increasing the likelihood of disease. More and cleaner water would help to reduce the incidence of intestinal infections and therefore of secondary malnutrition.

One of Africa's most pressing problems is associated with land management in the semi-arid zone of the Sahel – a fragile dryland ecosytem which cuts across Africa from the Red Sea to the Atlantic. In this region complex interactions between poverty, resource use and climate are exacerbating the fuelwood crisis and desertification, which, in turn, are allied to the region's widespread hunger. Politically charged debates about the causes of and cures for desertification have a long history (Adams, 1990). Their political character was most recently glaring obvious at the Rio Conference. The definition enshrined in Agenda 21 viewed land degradation in the Sahel as the result of several factors, 'including climatic and human activities'. The climatic element referred to the erratic rainfall in the region, specifically a 25 per cent decrease in average rainfall over the past thirty years, while the human activities included cutting down trees for firewood and unsuitable farming practices. African participants at the conference objected to this perspective because it failed to appreciate the wider context within which poverty in the Sahel must be understood, that is, historical colonial relations, misguided aid policies, falling commodity prices and the pressures of debt repayments. Unfortunately, paralleling all the heated debates has been an intensification of the symptoms of environmental crisis, and its human costs. A variety of efforts by non-government organisations as well as governments have been initiated to try to reverse the environmental degradation.

The food security role of livestock in a remote rural area may be critical. In arid and semi-arid regions, where cultivation is limited, people's entitlements are often based on a variety of strategies (Box 6.1); one of the most important is their ownership of livestock. Livelihoods may be based on nomadic patterns, which ensure that animals have access to water and pasture throughout the year. Livestock are relatively drought-resistant compared with crops and they are an extremely flexible asset; this flexibility is crucial to the survival strategies of rural populations in semi-arid areas. Animals, usually cattle, camels, sheep and goats, may provide food; they supply energy for agriculture in the form of their draught power or manure; they are a form of capital which may be allowed to appreciate (grow and breed) or be realised for cash when required. In areas where rainfall is unreliable, another characteristic is vital – animals are mobile and can be herded towards new pasture and water sources in drought conditions.

Although communities may survive a one-year drought, a pernicious interaction occurs in semi-arid regions during a severe or prolonged drought between animal prices and grain prices. Local livestock and grain markets interact to reduce income and increase food prices simultaneously. The mechanism is deadly and simple. As a drought intensifies, grain scarcity forces up grain prices; at the same time herders find it difficult to find water to sustain their animals and need more cash to buy food at the higher prices. As more

Box 6.1

---

**Mali**

Since independence, Mali has experienced frequent food crises. Famine occurred between 1968 and 1973, and lesser food crises existed in the droughts of 1983–84 and 1988. In addition to periods of acute food shortages, seasonal food insecurity is widespread. Seasonal food shortages occur in the rainy season prior to the harvest, when cereal supplies are low but energy needs related to agricultural work are at their highest. The rainy season is from July to October, and rainfall is 600–800 mm. The Bambara agriculturalists' major crop is dryland millet, and they usually keep some animals; some income is raised from the marketing of groundnuts, crafts and petty commerce.

Rainfall was very poor in 1988 and only 11 per cent of households harvested enough millet to satisfy their domestic needs; 53 per cent of households were required to ration food intake and substitute wild foods in place of cereals. In this precarious environment, survival depends on a household's ability to minimise risk. A range of strategies are employed to enhance subsistence security and to cope with food crises when they occur. Cropping strategies such as early-ripening varieties, field dispersal, intercropping and staggered planting help to spread production risk in time and space, while market gardening, craft production, migration and livestock investment permit accumulation of assets to realise in times of need. There are an interesting array of non-market strategies and social institutions such as collective labour groups, communal fields, village cereal banks and kinship exchange relationships, which provide a form of welfare or subsistence guarantee to participant households. Networks within and beyond the village are important in times of stress.

*Source*: Adams (1993, 41–51).

---

and more animals are sold, their prices fall, so generating a scissors effect – high food prices and falling assets. The extreme outcome is widespread famine. While attention needs to be drawn to some of these connections, it is none the less important to reiterate that the role of climate must be conceptualised as a proximate variable which has limited explanatory power compared with the political and socio-economic context within which the communities 'at risk' from drought are required to live.

**Plate 6.3** Animal market, Sokoto, Nigeria

**Plate 6.4** Fulani women transporting milk to market
**Plates 6.3 and 6.4** Livestock and entitlements
Large livestock, as shown in Plate 6.3, as well as small livestock production (pigs, guinea pigs, chickens, rabbits, etc.) is often a crucial component of rural and urban livelihoods. In dry areas, livestock are particularly important and contribute in numerous ways to survival; large livestock provide draught power, transport, manure (for fertiliser and fuel), milk, meat and hides.

Unfortunately prices for livestock can vary greatly and in drought the price of animals plummets as people sell them off to realise cash; at the same time cereal prices increase.
*Photos*: Dr Hamish Main, Staffordshire University.

## Ethnic and religious minorities

Other sub-national distinctions are relevant. Minority groups, ethnic or religious, often suffer a greater incidence of poverty and hunger. Where disadvantaged regions coincide with ethnic or religious minorities, problems are intensified and regional disadvantage becomes synonymous with ethnic or religious discrimination. Religious and ethnic communities identified as particularly prone to malnutrition are numerous and are found in both the North and South.

Peasants in China before the twentieth century frequently suffered from catastrophic famines. Since the revolution in 1949, with one very notable exception, the famine in 1959–61, famine has largely disappeared. Poverty and malnutrition persist, however, especially among specific vulnerable groups. Despite the phenomenal economic expansion since the 1978 reforms and improvements in the circumstances of a great many of its people, some contemporary analysts accept that economic expansion has increased polarisation between rich and poor and that there are strong regional and ethnic dimensions to the pattern of poverty (see Box 4.1). Some stark regional contrasts are obvious in Figure 6.2.

## Regional contrasts in China

A high proportion of the absolute poor are concentrated in several north-western, northeastern and southwestern provinces, where ethnic minorities are also concentrated. In 1989, it was estimated that at least 10 per cent of the rural population of the northwestern provinces of Shanxi, Henan, Shaanxi, Nei Meng, Ningxia, Gansu, Qinghai and Xinjiang, the northeastern provinces of Jilin and Heilongjiang and the southwestern provinces of Guizhou, Guangxi and Yunnan were 'absolutely poor'. The absolute poor also accounted for 6 per cent of the population of Liaoning and 7 per cent of Sichuan. These fifteen provinces account for approximately 80 per cent of the Chinese population categorised as poor. They are concentrated in deep mountainous regions with poor natural resources, minority nationality populations, endemic diseases and infrastructural deficiencies, including serious shortages of clean water. The government has instituted poverty alleviation programmes, but a recent analyst is not confident of their effectiveness (Croll, 1994). The following description confirms that China is indeed a country of stark contrasts (see Box 4.1):

In Canton the national obsession with food ascends to a guzzling crescendo. At night the porticoed pavements become a chiaroscuro of celebrants munching snacks or hunting down eating-places. Above them half the neon signs dangle fused, but a crossfire of pop music blazes from stereo shops, and hilarity engulfs the food stalls. The restaurants rise in multi-tiered pagodas – bursting palaces of merriment and greed. A trinity of statued gods presides in the reception-halls, where lanterns and gilded pillars glimmer fatly under chandeliers like inverted lotus-blooms. On the lower floors the feasters assemble in parties of ten, twelve or sixteen – often all men. Their tables fill up with dishes which everyone shares – sea cucumber, silver fungus, water chestnuts in tomato puree, beancurd, abalone soup, giant prawns, toffeed sweet potato. The expectation is unbearable. Every course drops into a gloating circumference of famished stares and rapt cries. Table manners joyfully endorse the ecstasy of eating. Diners burp and smack their lips in hoggish celebration. Bones are spat out in summary showers. Noodles disappear with a sybaritic slurping, and rice bowls ascend to ravenously distended lips until their contents have been shovelled in with a lightning twirl of chopsticks. The host heaps up the bowls of the guests on either side. Somebody else rotates the laden turntable in the centre, while the banqueters reach out and lever up a sugared walnut here, a lotus root there.

(Thubron, 1988, 182)

Two other distinct, sub-national, categories of people who often suffer from food insecurity and hunger are displaced people and refugees. Because their plight is usually related to conflict, they are considered in the next chapter.

## Households and food security

Where the incidence of poverty increases, it is principally among rural and urban employees – especially among resource-poor and often female-headed households, living in resource-poor rural environments or on the margins of urban agglomerations.

(Lipton and Maxwell, 1992, 3)

The last scale examined is food security at the local level and it focuses on the household and its individual members. This topic was introduced in Chapter 5, where the discrepancies between male and female entitlements were considered; here other aspects are considered. Local food security is structured by factors in the international and national arenas, but there are significant variables which help to explain contrasts within even quite small

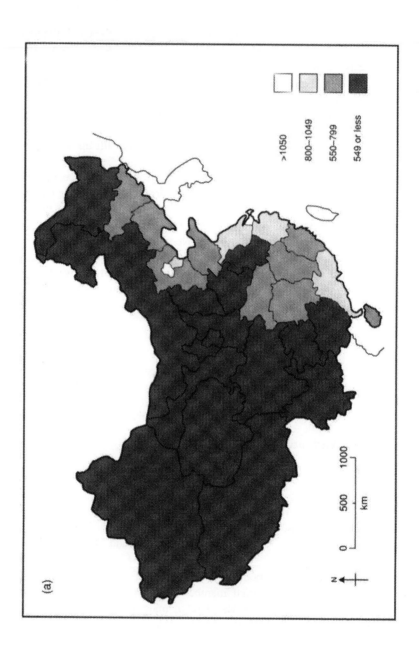

(a)

>1050

800–1049

550–799

549 or less

0        500        1000
         km

N

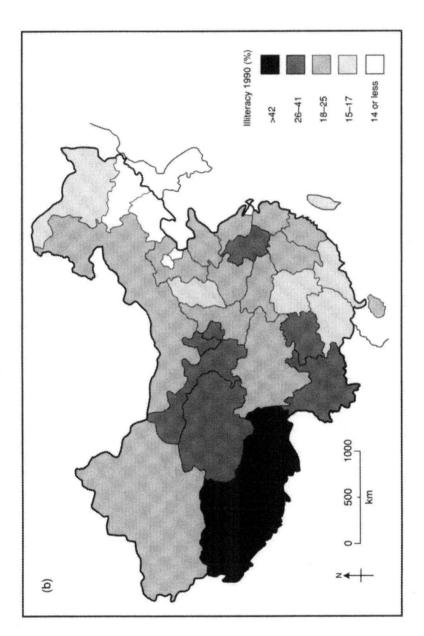

Illiteracy 1990 (%)

- >42
- 26–41
- 18–25
- 15–17
- 14 or less

(b)

**Figure 6.2** (a) Annual rural incomes in China, 1989 (yuan per capita); (b) Illiteracy in China, 1990

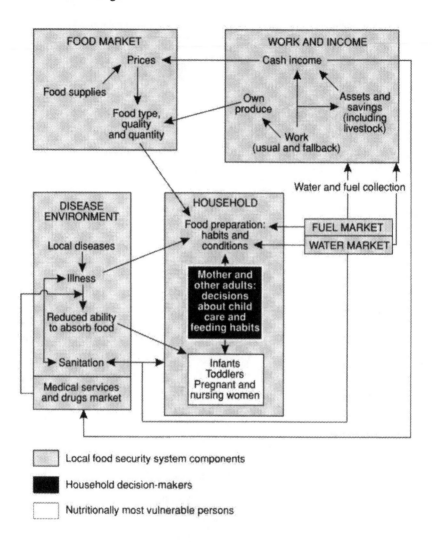

**Figure 6.3** The local food security system
*Source*: Hubbard (1995, 8)

areas. Figure 6.3 (2.1 in Hubbard, 1995, 8) illustrates some of the important interactive elements. Changes in one or more of these will have repercussions for household food security. The diagram shows how important variables in the household may be to local food security. The characteristics of the household, in turn, mediate the impact of these external changes and its members will experience the changes in different ways, depending on their role and status within the household. These local food systems are dynamic and open; they are always adjusting and may change drastically because of alterations in external or internal factors, naturally or socially derived traumas or benefits – floods, crop failure, increased tourism, non-government organisation (NGO) or state interventions.

It is possible to distinguish degrees of vulnerability between the poorest households. Chambers noted that in the developing world 'It is households that are increasingly distinct economic entities for production, for earning, and for sharing consumption.' It is important therefore to attempt to understand why some households may be more susceptible to hunger than others. Such analysis is essential to inform effective policy interventions.

Households may be poor or ultra-poor (Chambers, 1983). Ultra-poor households eat below their estimated weight-adjusted dietary requirements, despite spending at least 80 per cent of their income on food. One estimate is that in developing countries 2–5 per cent of households are in this category but that as many as 10–20 per cent of households may be at risk of falling into it. What characteristics do these poor households share and what happens to make the poor slip into the ultra-poor category? Chambers lists five clusters of disadvantage:

- *The household is poor*, assets are few, it owns no land or a small marginal holding. Indebtedness is a problem and stocks and flows of food and cash are low and unreliable, seasonal and/or inadequate. Strategies for survival are varied but tenuous;
- *The household is physically weak*, it has a high ratio of dependants to able-bodied adults and may be headed by a single adult.
- *The household is usually isolated*, having a peripheral location, it has limited access to information and members are often illiterate.
- *The household is vulnerable*, it has few buffers against contingencies, which means that any social/physical disaster results in increased poverty. It is often seasonally extremely vulnerable, if rains are late or food prices increase because of poor harvests or the diversion of food elsewhere.
- *The household is powerless*, its members are vulnerable to extreme exploitation and are in a weak negotiating position.

These characteristics help to explain why these households have few and tenuous entitlements and why some or all of their members may suffer from constant or occasional malnutrition. Given their structural vulnerability any number of proximate variables may impoverish the household further and cause a poor household to become ultra-poor. Again Chambers' (1983) work is helpful. He identifies five types of condition which may initiate a spiralling downward of a poor household's fortunes:

• social conventions which demand that the household spends money – dowries, bridewealth, weddings, funerals;
• disasters, which may be social in origin (theft, vandalism, war, persecution), or natural in type (floods, droughts, epidemics, crop failures);
• the physical incapacity of one member, a prolonged illness, pregnancy, childbirth, accidents, may result in the household moving from the poor to the ultra-poor category;
• unproductive expenditure can erode their limited assets – a failure in business, litigation, gambling, drugs;
• these families are particularly vulnerable to exploitation through exorbitant interest rate charges, corruption or bad deals.

It is important to recognise that *household* survival strategies are employed when food deficits exist. When a food shortage occurs households have a variety of options open to them. Decisions are made with an appreciation of the long-term costs and consequences of each option. A household may opt to reduce consumption rather than sell assets to maintain consumption, so that malnutrition may be a temporary manifestation of a temporary strategic household decision. Understanding the specific contexts within which households make these decisions is very important to the formulation of policies designed to reduce hunger; interventions should work to maximise the household's long-term survival.

## Trees and household entitlements

'When we were young we used to go to the forest early in the morning. . . . In a short while we would gather all the fodder we needed, rest under the shade of some huge tree and then go home. Now, with the going of the trees, everything else has gone too' (quoted in Sontheimer, 1991). It is difficult for people living in the North to appreciate the role of trees in rural household economies in the South. Trees may be of direct relevance to food security and variety (Box 6.2). In the home gardens of Southeast Asia, trees are grown to provide fruit, for example bananas, coconuts, sugar apples, mangoes, star apples, guavas, avocados and breadfruit. In some regions food from trees is vital as a safety buffer in times when harvests are poor: in Tanzania, two or three tree species supply some food for every month of the year and research has revealed that they are vital buffers in times of famine and/or drought.

Box 6.2

### Sri Lankan case study: the jak and breadfruit trees

There is a growing problem of undernutrition in Sri Lanka, which is particularly serious for the most vulnerable groups. Compared with the 1980–82 figures, the 1987 nutritional surveys showed a marked increase in chronic undernutrition in the age group 6–35 months. Problems of food security have been aggravated by the decline in real incomes following the structural adjustment programmes. However, research has suggested that large quantities of a potential food source are not usefully exploited. It is estimated that approximately 55–60 per cent of indigenous fruits are wasted every year. This situation has generated efforts to find ways to improve utilisation rates of these fruits. The two most important sources of wasted food are from trees, the jak and the breadfruit.

The jak tree is indigenous to Sri Lanka and grows in most parts of the country. Its fruit is used for food, the bark for timber, leaves for animal fodder and the dead wood for fuel. Average tree height is approximately 25 feet, and it begins to bear fruit in its third year. Depending on the tree variety, fruit is harvested once or twice a year. The breadfruit tree is not as common as the jak tree, but it is also very popular. The tree grows to a height of 15–20 feet and is popular among all income levels and with rural and urban families. Both trees are grown in home gardens, and are collected from common lands or from government reservation lands near homes.

In addition to being eaten as fresh fruit, jak and breadfruit are processed by female family members into secondary products by sun-drying. For rural women, these dried products improve food security. The dried fruits can be kept to replace or supplement the staple food crops – rice, yams and wheat flour – when these are scarce or expensive. During the rainy season, these are important dietary standbys because demand for casual labour is low (therefore cash is scarce), and the distribution of staple foods is erratic in very wet weather. These foods

**Plates 6.5–6.7** Trees and entitlements

Trees are obviously sometimes important sources of food (Plate 6.5), but less directly they impact entitlements by providing shade for animals and people and by reducing evaporation in arid areas (Plates 6.6 and 6.7). Deforestation and its implications for local sustainability continue to cause concern in many parts of the South.

*Photos*: Dr Hamish Main, Staffordshire University.

**Plate 6.5**  Guava tree, Kano state, northern Nigeria

**Plate 6.6**  A village on the Jos Plateau, Nigeria

**Plate 6.7**  Kebili Oasis, Tunisia

Box 6.2 (*continued*)

have another useful characteristic: they are filling and help to reduce or postpone hunger pangs. Processed fruit is used primarily for household consumption, but some surplus is sold or used to fulfil social obligations, including almsgiving and gifts to the temple. Although primarily used within the household, some processed fruit is funnelled through informal and more formal economic links. Large surpluses are sold through small traders and retailers to be sold to urban populations, while prepared snacks are also sold through informal trading networks.

Despite all these advantages, much of the fruit is neither processed nor consumed fresh but is wasted. While this appears irrational in a context of food insecurity it must be viewed as a rational response to several constraints. One constraint is the time-consuming character of the processing task and a second is that even low wages, available as casual labour, are more beneficial to household survival than fruit processing. Techniques available for processing the fruits remain basic. Certainly there is an opportunity for interventions to reduce the time required for processing, thereby making it more economically viable. In addition, interventions could improve the market prospects for the products as well as the percentage of the profits enjoyed by the producers rather than the traders.

*Source*: Abeywardene (1995).

In many countries in the South, wood, gathered locally, is the dominant source of domestic energy in rural *and* urban homes. Collecting fuelwood for domestic consumption is usually a female task, and in recent years it has become much more onerous. As the pressure on local supplies has increased, women must walk further from the village to obtain sufficient supplies – this is one manifestation of the fuelwood crisis. Fuel is essential for cooking so ample supplies must be gathered, but as women are forced to devote more time to collecting wood, other essential reproductive activities get neglected.

Trees are an important source of animal fodder in many countries and at certain times of the year they are critical for animal survival. In most of the rural South, animals are a vital element of household entitlements; poultry, pigs, goats, rabbits, guinea pigs as well as the common larger livestock are flexible buffers between households and hunger. This is not limited to rural areas; in many urban households livestock are kept to supplement household diets and income. In Nepal, women are responsible for finding fodder for the buffalo, which are stall-fed and essential capital for households. One buffalo requires up to 40 tonnes of grass and leaves a year; feeding them is one of a woman's most demanding contributions to household survival.

Trees provide the raw material for numerous other income-earning household-based activities. Among the most important from different areas are canes for furniture; fibre for nets, ropes and mats; bamboo for basketry and construction materials; gums and resins; leaves for making cigarette wrappings, estimated to employ 600,000 women and children in India; wood crafts for sale. Supplying charcoal to the urban market is an important source of income for many rural people.

Shade is essential, for people and animals, in many hot climates; again trees are useful. They supply medicines and dyes, and help to increase soil fertility and reduce erosion, especially in fragile semi-arid ecosystems. In addition, trees provide the raw material to build shelters and make utensils. This brief overview begins to illustrate how vital a reliable and diverse source of wood is to the survival strategies of poor households. If trees become scarce or less accessible because of changes in land ownership, poor households are forced to spend money to purchase the fuel, food, fodder or timber they require. For all these reasons the depletion of forest resources is very serious and has negative implications for the day-to-day material circumstances of poor families.

The mapping of deforestation is difficult but in most parts of the South it is certainly a reality that the poor do not dispute. The problem of fuelwood depletion has immediate implications for diets. As fuelwood becomes scarce, poorer wood is used, which makes cooking more difficult, and it may alter cooking methods. Sometimes women decide to cook less frequently, so a cooked meal may be offered only once instead of twice or three times daily; this modification has been recorded in parts of West Africa and in the Andes. Traditional foods which require prolonged cooking may be replaced by more raw foods. Research is just beginning to reveal the implications these changes have for nutritional standards, and they are not positive.

## Conclusion

This chapter reviewed distinctions at the sub-national level which influence entitlements. All countries have contrasts within their boundaries but in some cases the divisions are very stark indeed. Regional diversity may be based on inherited patterns of disadvantage and in most countries minorities are more likely to suffer malnutrition than the majority ethnic group. Some household types are more prone to food insecurity than others and some household members have a higher incidence of hunger than others. Where social and economic change is rapid, new patterns of vulnerability emerge, sometimes exacerbating existing inequalities and sometimes creating a new group of disadvantaged people. Sub-national patterns of hunger are dynamic and can be transformed for better or worse by political interventions, or by economic or technological changes.

## Key ideas

1 Contrasts in entitlements within countries can be very marked, by region, ethnic group, caste, age, gender, class or between rural and urban areas.
2 Some regions within countries have long histories of disadvantage *vis-à-vis* other regions.
3 Minorities are more likely to suffer from hunger than majority populations.
4 Economic success at the national level is not necessarily reflected in improvements in the nutritional status of all regions or groups of people within a country; sometimes quite the opposite happens, i.e., economic expansion may intensify regional or ethnic disparities.
5 People's entitlements may vary significantly by season; this is especially crucial for poor households with limited resources.
6 Climatic variables may interact with social variables to create areas of famine or chronic food insecurity.

# Chile

## PASTEL DE CHOCLO (Corn/maize pie with chicken)      Serves 4–6

Variations of this pie are found in Bolivia and other Latin American countries. Corn or maize is the only major cereal grain native to the Americas. In North America, it was introduced to the colonists by the Iroquois Indians. Among the many varieties is the distinctively flavoured blue-corn, grown in Arizona and New Mexico by the Hopi Indians. For this recipe you can also use left-over cooked chicken, and miss out the first two steps in the method.

## INGREDIENTS

3 lb/1.5kg chicken, skinned and cut into portions

| | |
|---|---|
| ½ cup/75g olives, cut in halves | 1 medium can corn kernels |
| 2 hard-boiled eggs, sliced* | 1 teaspoon dried marjoram |
| 2 tablespoons oil | 1 teaspoon sugar* |
| 2 eggs, beaten | 1 tablespoon flour |
| 2 onions, chopped | salt and pepper |
| ¾ cup/180ml milk | 1¼ cups/300ml stock or water |
| 1 teaspoon ground cumin | ½ cup/50g raisins or sultanas |

* optional ingredient

## METHOD

1 First place the chicken portions into a large pan and pour in just enough water to cover them. Put a lid on the pan and bring it to the boil. Then lower the heat and simmer for 30–40 minutes or until the chicken is cooked.

2 Lift out the chicken pieces, keeping the stock, and allow them to cool. Then remove the meat from the bones.

3 Now heat the oil in a pan and sauté the onions for a few minutes before adding the cumin and marjoram. Cook for 2 minutes before sieving in the flour, a little at a time.

4 After that, slowly add the stock or water, stirring constantly so that you get a smooth sauce. Bring this to the boil and then put in the chicken pieces, seasoning with salt and pepper.

5 Next, spoon the chicken mixture into an oven-proof dish and scatter the raisins or sultanas, sliced olives and hard-boiled eggs on top.

6 Then whisk the beaten eggs and milk together in a bowl, adding the corn. Pour this over the chicken and sprinkle the sugar on top. Bake in the oven for 45 minutes or until the topping is set. Serve with baked potatoes and salad.

# 7
# Conflict and hunger

## Introduction

This topic of this chapter, like Chapter 5, cannot be analysed with reference to one scale. Conflict as experienced in most parts of the world today is caused by factors at the international, national and local levels. It tends to become most serious in places where resources are already stretched, which itself may be part of the causation, and inevitably its consequences mean a reduction in these already inadequate resources. Its impacts are also experienced, in different ways, at all levels and of course at the household level too. Its impact is also differentiated by gender, age, ethnicity and religion in most cases. Unfortunately, the implications of conflict for food security remain very serious: it is particularly implicated in the creation of famine circumstances. Analysis of entitlements, broadly conceived, around which this text is built, makes the examination of conflict unavoidable; conflict erodes, destroys, despoils and transforms the entitlement patterns of all those caught in its scope.

> The old battlefields of the Third World, where the proxy wars of the United States and USSR were played out, are still bloodied by conflict in countries such as Afghanistan and Angola. In other countries, new waves of violence are accompanying the birth of a post-Cold War (dis)order.
>
> (Macrae and Zwi, 1994, 1)

Today (23 July 1996), Burundi and Rwanda are in the news again and the international community holds its breath and hopes that more massacres will be avoided – holds its breath because it appears incapable and/or unwilling to do anything more constructive. This situation, in East Africa, is just one example of what is referred to in academic, political and policy circles as a 'complex emergency'. These precipitate humanitarian crises and are multi-causal but always associated with armed conflict. The number of complex emergencies is increasing and they are intimately and inevitably associated with food crises of varying types and intensity. The term 'food wars' (Messer, 1994) is applied to many of these conflicts and reflects the centrality of food to their dynamics. Although they occur most frequently and disastrously across the South, since 1989 they have occurred in the area of the former Soviet Union and with devastating humanitarian consequences in the former Yugoslavia. However, their gravest direct and indirect impact is in sub-Saharan Africa. Measuring the number of deaths due to conflict is problematic, but Green holds that in sub-Saharan Africa since 1980 'much below eight million would be unrealistic' (Green, 1994, 37).

As the name implies, these emergencies are caused by several interacting factors, and they always precipitate food crises and associated high rates of morbidity and mortality. This chapter examines some of the ways that armed conflict and hunger are related. Although, as emphasised in Chapter 2, deaths from famines are fewer than deaths associated with chronic hunger, famine deaths nevertheless still occur, and are most frequently recorded in conflict situations, because humanitarian relief is diverted, appropriated or obstructed. Chronic hunger and malnutrition are also found in conflict situations. Even when the conflict is meant to be over, the legacy of conflict may prevent people from accessing a basic diet for years; rehabilitation of devastated socio-economic communities requires time, money and political will; these may not be forthcoming locally, nationally or internationally.

## Conflict and hunger: the connections

Food crises associated with conflict need careful analysis and each case displays distinct characteristics, but they all illustrate the complex nature of hunger, its creation and politics. International humanitarian relief efforts in these situations have proved daunting and, in some circumstances, doomed (Somalia, Bosnia, Ethiopia, the former Yugoslavia, Afghanistan). Convincing analyses of complex emergencies require an appreciation of the broad historical and structural circumstances within which they occur as well as recent global processes that have intensified economic, environmental and social processes of marginalisation in specific regions and localities and among specific ethnic groups. Without exception this is the case. Consider the following examples from different world regions and the limitations of adopting a too narrow spatial or temporal perspective: Afghanistan; Cambodia; Angola; Sri Lanka; Ethiopia; Iraq; Tajikistan; Croatia; Guatemala. In all these examples, the legacy of colonialism (all manner of boundaries, real and imagined) and the Cold War (superpower rivalry, i.e. the geopolitical context), the regional ramifications and repercussions of the conflicts, are different but essential to understanding the specific contours of these contemporary tragedies.

In another important respect these emergencies confirm a major argument of this text, that there is seldom much that is accidental, inevitable or 'natural' about hunger and famine in the contemporary world. If these emergencies were 'natural' then relief efforts would be simple and successful – 'we have the technology'; they are never simple and rarely as successful as they might be because these disasters are complex and politically inspired. Naive international interventions, based on simplistic understandings of the situation, may occasionally exacerbate the crisis rather than relieve its victims. That is not to suggest that interventions are never useful but to warn that, on the basis of recent experience, they may become part of the problem:

NGOs have played an important role in mitigating the effects of humanitarian crises accompanying conflict, often enabling the survival of significant numbers of people. However, evidence from the Horn and southern Africa suggests that their interventions, supported by international donors, have had significant and unpredictable side effects: relief operations have often contributed to the conflict dynamic, and supported the growth of war economies.

(Duffield, 1994, 227)

When we read about the 'crisis' in Burundi or Afghanistan we too easily assume that the people who live in these places are all suffering; in fact, this is seldom the case. It is becoming increasingly clear that these complex emergencies serve a purpose; there are beneficiaries as well as victims of these conflicts and their associated food crises and dislocations. Many of these situations become self-generating because vested interests emerge which have a stake in continued conflict. Fragile political regimes, often with limited legitimacy among large segments of the populations they govern, may have little to lose and a lot to gain from famine and its attendant social dislocations. Oppositional forces also operate in a 'conflict space' (decline in civil institutions, breakdown in official accountability and law and order), which offers opportunities as well as costs. For the principal antagonists then, profiteering may become the fuel of conflict rather than ideology or injustices.

Food crises, and their occasional most serious final sequence, famine, may be the political *objective* in some cases. Analyses of recent cases confirm that there are beneficiaries of the crisis, who maintain a volatile coalition to exploit the spoils of conflict. Both government forces and oppositional forces require food to sustain their armies and money to sustain their military campaigns. In a complex emergency, there are large inputs of food aid and relief supplies, and relatively massive injections of cash to distribute food and relief services; either may be diverted from their intended targets and end up sustaining military operations. In most conflict situations, parallel economies emerge which are impossible to quantify but are judged by experts to be very lucrative. Apart from the obvious ones alluded to above, syphoning off relief supplies of food and cash, there are other less obvious ways to exploit the situation. These alternative modes of income generation can be understood as entitlement packages which are, if not unique to conflict situations, certainly more important in such circumstances. Armed people, rebel or government, have a certain power which those without arms have not. They can usually 'negotiate' very favourable 'terms of trade', for food, shelter and gifts to support their cause. The 'war economies' which are created may dwarf the formal economy which prevailed before the conflict or which is likely to emerge from its ashes. In sum, conflict situations often create novel forms of entitlement for *some members of a community*; sustained by violence or illegality, these may be well worth the risk for an otherwise impoverished population. As noted, 'in many parts of the world, the prospects for a lasting peace depend on weaning various groups away from violent strategies that have, for them, served important economic as well as political purposes' (Keen and Wilson, 1994, 216).

## Conflict and hunger: direct impacts

The direct impacts of war or conflict are obvious and massive in some areas. Scorched-earth tactics effectively destroy all agricultural production and have been used most recently in the Horn of Africa. Essential agricultural infrastructure, irrigation canals, grain stores, agricultural support services and stocks (veterinary services, fertilisers and seeds, diesel, extension advice) are all easy targets which have immediate consequences for food production. Another strategy available to the combatants is to pepper the rural areas with land mines. The impact of these is horrific and concentrated in the civilian community: 10 per cent of agricultural land in Eritrea is mined; Afghanistan is the most heavily mined country in the world; 15,000–40,000 people in Angola are mine-related amputees; over 1 million mines have been planted in Somalia, which means that large tracts will be dangerous for years; in Cambodia there are two mines for every child, one person in every 236 is an amputee, and mines claim over 500 victims a month (Watkins, 1995, 48; Macrae and Zwi, 1994, 15). Animals and grain stores may be confiscated, appropriated or burned. In Somalia and Sudan, pastoralists found wells, essential for watering their animals, poisoned or destroyed.

A direct loss of entitlements, in the shape of losing land ownership or access, is experienced by many in conflict situations; massive out-migrations are one of their most characteristic outcomes. People may be forced out by various methods – fear and intimidation, evictions in ethnic cleansing operations, blockades of supplies, a variety of ways – but the outcome is massive farm abandonment. Often, however, the vacuums so created are short-lived and other populations, more acceptable to the armed elements, settle the vacated properties. Such asset transfer may become synonymous with cultural genocide and Duffield is persuaded that this process is well underway in Sudan as far as the following peoples are concerned: Dinka, Fur, Nuba, Maban, Uduk, Anuak, Chai, Murle, Toposa and Mundari (Duffield, 1994, 55). Allied circumstances exist in Bosnia, Angola, Mozambique, Rwanda, Somalia, Zaire, Afghanistan, Guatemala, Cambodia, Burma and elsewhere. In this situation too, there are winners and losers. Other basic assets may be transferred to the detriment of some and the benefit of others. Emergency sales of cattle are common in conflict situations where famine looms. Cattle may also simply be stolen and transferred. Evidence from Sudan is pertinent again: Duffield estimates that as many as 340,000 cattle were stolen from communities in southern Sudan and ended up being owned by merchants and traders in the north of the country between 1984 and 1986.

Another common misconception about conflict situations is that most of the casualties are soldiers and military personnel or interested parties closely allied to the combatants. In most recent conflict situations, the people who bear much of the costs are the civilian population, who 'are threatened not only by bullets and bombs, but by the massive social and economic dislocation engendered by war. War undermines those very social, economic and political systems upon which we rely to secure our basic needs, including the most basic – food' (Macrae and Zwi, 1994, 1). Oxfam estimates that four out of five casualties of conflict today are civilians, most of them women and children (Watkins, 1995, 43).

Equally serious for urban and rural residents and their food security is the destruction or besieging of cities, ports, airports and markets. The availability of vital imports ceases; most of these countries are not self-sufficient in food even in good times and the absence of other basic raw materials has ramifications for all economic sectors. Access to basic health services is eroded and essential health products become unavailable. All medical facilities are stretched attending the wounded and trying to delay the appearance of the diseases associated with such circumstances: diarrhoea, dysentery, hepatitis, cholera, malaria, typhus, measles and respiratory diseases. As discussed in Chapter 2, the relationship between malnutrition and disease is complex; when a city is blockaded, food is scarce and disease is rife, and these circumstances are mutually reinforcing. De Waal (1993) cites the siege of Juba in southern Sudan as the longest-running one in contemporary Africa. Between 1984 and 1992, the Sudan People's Liberation Army managed to block supplies to the city by a variety of methods: a ring of soldiers, encircling by land mines, attacking road convoys and shooting down relief aircraft.

Movements by people and animals may be severely curtailed during conflicts and this can have very serious consequences for food availability, especially in crisis circumstances. One of the first indicators of a food crisis is that people migrate in search of food, work and markets where these may be available and where they might sell some of their assets to exchange for food. Losing access to such markets because travel is disrupted, animals sold, roads and bridges destroyed, or because blockades exist, or because people simply fear what might become of them at their destination, means people become destitute more rapidly.

So restricting people's mobility may exacerbate their plight but so too, ironically, can the forced migration of people in conflict situations. Government counter-insurgency campaigns, their attempts to destroy opposition forces, often result in massive relocation of civilian populations to protected villages or secure zones; such strategies have a long history in warfare. De Waal (1993) describes how these tactics were employed by different regimes in Ethiopia in the 1960s and 1980s. He also cites the policies of the Mozambique government in the 1980s in Zambezia province; this 'villagisation' effort he holds largely responsible for the creation of a severe famine in that area in 1986–87. These forced relocations rip whole communities away from the resources essential to their survival, not only farm land, crops and animals, but from places they know intimately, places they can exploit successfully in times of hardship, the location of wild foods, fuelwood and water resources in drought years. The rhetoric used to justify resettlement policies may sound convincing – 'for their own good' – but in fact may be a completely hollow and cynical 'we want them out'. Other massive forced migrations with consequences for entitlements and food security are reviewed in Box 7.1.

Box 7.1

## Refugees and displaced people

In 1993, approximately 23.5 million people were recognised as refugees by the United Nations High Commissioner for Refugees. The great majority of these people are refugees from conflict and environmental collapse. All major world regions are both sources and hosts of refugees and it is a very changeable pattern. In 1988, for example, most lived in Asia (6.8 million) while by 1993 most lived in Europe and the former Soviet Union (8.3 million). The figures for Africa fluctuate but remain high, 4.6 million in 1988 and 7.5 million in 1993. About 80 per cent of these uprooted people are women and children; many live in women-headed households. The definition of a refugee is enshrined in the United Nations 1951 Convention and 1967 Protocol Relating to the Status of Refugees as one 'who could not return to their country of origin because of the *well-founded fear of persecution* on account of their race, religion, nationality, political views, or membership in a social group'. The ambiguity of the term 'well-founded' has been a major source of debate. When does discrimination become persecution? The onus is upon those seeking to escape persecution to convince the decision-makers in the host country of the validity of their case. Another confusion arises because, in a world political economy characterised by gross structural in-equalities, the distinction between economic and political refugees becomes problematic. This helps to explain why the Organisation for African Unity (OAU) has developed a regional complement to the UN's definition to include population movements caused by external aggres-sion, occupation, foreign domination or events seriously disturbing public order. Although an improvement, even this extended definition is unsatisfactory.

The term 'refugee' can be applied only to people who have crossed an international boundary, yet there are even more displaced people, people who have migrated *within* national boundaries. Although they share all the problems associated with forced uprooting and refugees – loss of all entitlements (land, capital, employment, kinship networks, public relief) – they have no entitlement to international protection or relief. That is reserved for refugees as defined by the original United Nations Con-vention. In 1993, there were more displaced people than refugees, 26 million, and that is almost certainly an underestimate.

Box 7.1 (*continued*)

For all refugees and displaced people, becoming uprooted causes major changes in their lives. The impact on women and children is particularly poignant and, in some cases, traumatic, particularly when rape and sexual abuse become commonplace.

(Forbes-Martin, 1992, 5)

Unfortunately, numbers of people in both of these categories are growing and in some places have already reached crisis proportions. In sub-Saharan Africa, the situation is particularly acute where already stretched national budgets are required to sustain influxes of hundreds of refugees or dispossessed people (most recently in Burundi). Refugees and displaced people may end up in transit camps or permanent camps or cities, be resettled in an alien landscape or one similar to that they left, or find their way to friends or relations in neighbouring areas or far afield. Their stories are diverse and not all tragic. For a majority of these populations, however, entitlements are absent or tenuous. If they are in internationally recognised refugee camps, their entitlements are based on the obligation of the international community to provide relief food (the United Nations High Commission for Refugees or the World Food Program), but food security for many of these populations is fragile, based on unsustainable local production or food aid which is inadequate (Black and Robinson, 1993; Forbes-Martin, 1992). A recent survey estimates that 10–15 per cent of refugee children under the age of five are malnourished, with rates in some countries as high as 30 per cent.

The influx of refugees or displaced people is usually unwelcome, because they are viewed with suspicion by the local host community, which in most cases is poor too. Refugees and displaced people leave disastrous conditions, but few of them arrive in places where conditions are satisfactory. They experience the health, food and sanitation problems found everywhere in the South, but these are compounded by the refugee or displaced experience: exhaustion, psychological trauma, family break-up, being in an alien environment and often living in overcrowded conditions. All this means that the incidence of infectious and malnutrition-related diseases (typhoid, dysentery, cholera, diarrhoea and respiratory diseases) is higher.

Box 7.1 (*continued*)

Refugees and displaced people are ingenious in the strategies they adopt to survive and prosper against the odds. Contrary to the 'dependent' image, most of these people survive because they engineer ways to do so themselves. If they receive apt assistance at the right time their efforts can be increased exponentially. Appropriate interventions may include providing access to income-earning opportunities by improving traditional skills or offering training in new ones; giving grants or loans to help people to produce for market; and interventions to improve agricultural production. Finally, refugee and displaced populations are a heterogeneous group, and hierarchies of power and exploitation exist within as well as outside the communities. Most essential may be to encourage the empowerment of various groups within these communities to ensure that exploitation from within and outside is addressed.

## Conflict and hunger: indirect impacts

Indirect costs of war have various ramifications for food production and people's entitlements too. One of the most serious could have been discussed in Chapter 4 when examining national perspectives and governments' commitment and capacity to secure food for their populations; governments' investment and expenditure decisions are at issue here. In addition to the economic costs of direct conflict-related damage, many governments involved in conflicts invest massive proportions of the gross national product in arms purchases and militarisation. Money spent on securing large well-equipped armies, or militias, is not available for investment in agricultural improvements, industrial expansion or any social or economic infrastructural improvements. In Ethiopia, in 1984, a famine year, 46 per cent of the government's budget was spent on buying armaments (Devereux, 1993, 149). The diversion of development resources occurs at the international level too. International exchanges are in arms sales rather than goods to promote long-term development, and today more international humanitarian funds are expended on relieving victims of conflict than are spent on initiating development programmes.

Productivity is also seriously compromised by the diversion of labour. In a very militarised situation the majority of male labour is exploited for conflict and this has several serious implications for productivity, especially in the rural sector and food production. The diversion of male labour represents a serious loss of entitlement to communities and households. Heavy maintenance work, tree felling, and construction of buildings and irrigation ditches are not necessarily the sole preserve of men, but any community's or household's capacity to invest in activities which pay dividends in the long term is compromised by the loss of the productive male population. Shortages of male labour may cause modifications of traditional gender roles in agricultural labour, but this can only ameliorate the problem, not eliminate the crisis of rural reproduction. Positive changes in the position of women in the community or household have been identified by some observers, but this is little consolation for traumatised communities and individuals. Demographic imbalances as a consequence of conflict can have implications for the long-term viability of communities, and their marriage and household labour arrangements. Many of the issues reviewed above are exemplified by the situation in Afghanistan, where the crisis is not yet resolved (Box 7.2).

Box 7.2

---

### Children sold by Afghan widows

By Ahmed Rashid in Kabul

Afghan war widows are selling their children in Kabul's bazaars to prevent them from starving to death, as a humanitarian and economic disaster threatens the whole region because of Afghanistan's continuing four-year civil war.

'Tens of thousands of Afghans could die of starvation next winter unless the West redoubles its peace efforts and humanitarian relief,' said a senior diplomat in Kabul. 'The black hole of the Afghan economy is also crippling the economies of Pakistan, Iran and the five Central Asian republics because of the increase in smuggling and the trade in weapons and heroin.' Inflation is 50 per cent a month in Kabul, bringing daily devaluation of the currency, the afghani, as under-funded international relief agencies battle against the odds to feed Afghanistan's population.

Daily life has become a fight to find enough of the grubby afghani notes, which are still printed in Russia, to pay for food. Although the shops stock goods smuggled in from Iran and Pakistan, few people have the money to buy them.

'For 400,000 people out of Kabul's 1.2 million population, this is a handout economy run by the UN,' said a Western aid official.

Last October, the UN appealed for $124 million (£81 million) in emergency contributions to Afghanistan from Western donors to cover

Box 7.2 (*continued*)

the following twelve months. So far it has received only $35.6 million. Western countries have not given a single dollar to UN physical infrastructure projects deemed essential to put the economy on its feet. At a children's playground set up by Save the Children in the Microyan housing complex, hundreds of Afghan boys and girls play grimly on the swings with all the signs of malnourishment: swollen bellies, sallow faces and stunted growth.

'The children's plight is the worse because the diet is so bad and the rockets fired by the opposition Taliban create death, trauma and uncertainty,' said Sofie Ellisussen, the director of Save the Children in Kabul. 'War widows go out to find work and have to lock their children in the bathroom for the whole day so that they avoid rockets. Other widows are selling their children to stop them from starving to death.'

Western relief agencies have set up bakeries to provide cheap *nan* bread for the most needy – the 25,000 families of war widows and 7,000 families headed by disabled men who can generate no income for their families.

'The daily diet of most people is just *nan* bread and tea,' said Bibi Nasiba, a local official of the UN's World Food Programme.

At the Mazang bakery, a widow, Bibi Zohra, recounts how her house was rocketed first in 1993 and then three months ago by the Taliban. The attacks first killed her husband and then her eldest son.

'Day by day the situation is worsening,' she said. 'We don't trust any of our leaders or neighbouring countries to bring peace. Only the UN can bring peace, by making a neutral government in Kabul.'

The crisis for the aid agencies is made worse by the failure of the warlords to carry out any relief work and their refusal to form a coalition government. President Burhanuddin Rabbani in Kabul, General Rashid Bostum in the north and the Taliban in the south are using their funds solely for their war effort.

The smuggling of food, fuel and consumer goods from Iran, Pakistan and Central Asia to Afghanistan is crippling these countries' economies too. In 1994–95, Pakistan is known to have lost nearly £500 million in customs revenue. The hauliers who smuggle food use the same routes and lorries to smuggle out heroin worth £1 billion.

*Source*: *Daily Telegraph* 11 June 1996.

## Conclusion

Despite the optimism that was felt by some at the end of the Cold War, evidence of any peace dividends is very scant. Either the legacy, or current lived reality, of war damage and dislocation still obstructs material improvements and food security in Central America (El Salvador, Guatemala, Honduras, Haiti and Cuba), parts of South America (especially Peru), large tracts of sub-Saharan Africa, especially the former front-line states and countries in the Horn, the Middle East (Palestine, Iraq), the Balkans, and numerous countries in Asia (Afghanistan, see Box 7.2, Sri Lanka, the Philippines, Kashmir, Burma, East Timor and countries in the Caucasus).

Most of these conflicts have their origins in history but are fuelled by contemporary poverty and injustice. International media attention to poverty and gross human rights abuses, and hence political responses, appears too late to be effective. As the Afghanistan case illustrates, international interest is often fleeting and motivated by national self-interest, not altruism. There are problems with international interventions too. Often they are 'band-aid' in nature and may exacerbate rather than ameliorate the problems. Despite the rhetoric to the contrary, even humanitarian interventions are politically inspired or compromised: some human rights abuses receive attention (Cuba), others do not (Indonesia); some hunger receives relief (Somalia), some does not (Iraq and Afghanistan).

It is impossible to arrive at easy explanations of conflict and it is equally difficult in some circumstances, by no means all, to assign blame. At each level of analysis different elements must be drawn into the picture; the final canvas loses clarity as colours and lines merge. To dream otherwise is dangerous. Even if the rights and wrongs appear relatively straightforward at the outset the massive intervention by outside agencies and money can itself alter the situation. The best way ahead must be to encourage early interventions to address widespread human rights abuses, because these allow the perpetrators to assume immunity from international condemnation and sanctions. If these were checked successfully, then the incidence of the ultimate human rights abuse, to be starved, might be reduced. As emphasised elsewhere in this text, being hungry reflects an individual's or group's political status. The devastating loss as of entitlements, by individuals, households, regions, ethnic groups and nations, which occur in the event of armed conflict are weighty arguments for developing peaceful conflict resolution strategies. Sadly that is not a revelatory comment. What must be sought are circumstances where people have something to lose from conflict and few have much to gain; that requires changes in the international political economy and its ideologies as well as the more obvious institution of more accountable, democratic regimes in the South.

**Key ideas**

1 Violent conflict has serious direct and indirect implications for entitlements and food security.
2 Most humanitarian emergencies are protracted and politically complex.
3 Civilians, not military personnel, are the prime victims in conflict situations.
4 Large-scale population movements associated with war and conflict increase the migrant and host populations' vulnerability to disease and hunger.
5 In all conflict situations there are beneficiaries as well as victims.
6 Manipulating food supplies is a vital strategy in conflict situations.

# India

Cumin, or *jeera*, the dried fruit of a parsley-related plant, is a common spice in Indian cookery. It looks like caraway but has a very different taste (caraway seeds are not really used in Indian cooking). Aside from flavouring curries, cumin is also used in India as an aid to digestion.

## INGREDIENTS
3 lbs/1.5kg chicken, skinned and cut into portions
4 tablespoons/50g margarine or ghee
2 medium onions, sliced
2 cloves garlic, crushed
½ cup/50g creamed coconut dissolved in ½ cup/120ml of the hot stock (see below)
1 teaspoon ground cumin
½ teaspoon chilli powder
2 whole cardamom pods
1 teaspoon ground cilantro/coriander
1 tablespoon curry powder
1¼ cups/300ml chicken stock
salt and pepper

## METHOD
1 Heat half the margarine or ghee in a pan and cook the chicken portions until they are brown on all sides.
2 Meanwhile, in another pan put the rest of the margarine or ghee and add the onions and garlic. Cook them for a few minutes until they are soft and then put in the coconut milk, cumin, chilli powder, cardamom seeds, cilantro/coriander, curry powder, salt and pepper and stir well. Cook this for 2–3 minutes over a medium heat.
3 Now spoon the mixture on to the chicken, pour in the remaining stock, cover, and bring to the boil. Reduce the heat and simmer for 40 minutes or until the chicken is cooked. Serve with rice or chapatis.

# 8
# Alternative futures

## Introduction

A comprehensive analysis of modern interventions that have reduced hunger would require a second book, but it is important to conclude this text with a review of some positive changes and optimistic scenarios. Understanding world hunger is, after all, only a worthwhile exercise if it suggests ways to eliminate it. This text has described the complexity of factors at a variety of levels which help to explain why some people in some places continue to suffer from hunger. That complexity, the hierarchies of interacting structural and proximate variables, means that universal 'answers' are unavailable. The thesis adopted in this text does suggest, however, the relevance of one generalisation. The relatively powerless lack food, so strategies that promote their ability *to make and implement decisions* about resources will enhance their access to food. Employing the concept of entitlements, we can say it is essential that the poor have both sufficient and secure entitlements. To that end, changes are required at all levels, from the international to the household. But such a statement is so sweeping that its insights are limited; it is more productive to consider how such alterations may be effected in specific places. If the spaces within which hunger occurs are social spaces then we must examine the spaces within which change can occur to eliminate hunger. This chapter reviews potential changes at the levels employed as a framework in the book – global, national, sub-national – before considering the relevance of several concepts which are important at all levels. First, however, another review of some statistics.

This book opened with an overview of the state of world hunger, which was a snap-shot of things as they are now. Here I describe how the situation has been changing in the last two decades. The picture is not all bad and it is important to recognise that progress has occurred in some areas; some positive changes have been discussed in various places in the text but it is worthwhile to consider general trends. The following list illustrates some of the positive improvements which have been made in the quality of life of some people in the South. Although not direct measures of nutrition these figures reflect an improvement in general circumstances, including nutrition.

- Between 1960 and 1992, average life expectancy increased by more than a third. Now thirty countries have achieved a life expectancy of more than seventy years.
- Over the past three decades, the proportions of the population with access to safe water has almost doubled, from 36 to 70 per cent.
- Net enrolment at the primary school level has increased by nearly two-thirds during the past thirty years, from 48 per cent in 1960 to 77 per cent in 1991.
- The combined primary and secondary school enrolment of girls has increased from 38 to 68 per cent during the past two decades.
- Despite rapid population growth, per capita food production has risen by more than 20 per cent during the past decade.
- Between 1960 and 1992, the infant mortality rate has more than halved, from 149 per thousand live births to seventy.
- During the past two decades, the lives of about 3 million children have been saved each year through the extension of basic immunisation.

(United Nations Development Programme, 1995, 16)

Despite this progress, real human deprivation remains serious in some places and populations.

The positive changes identified are not simple to interpret. Changes which reduce the incidence of these variables and hunger are the result of a number of interacting elements but it is possible to suggest what have been the most significant changes. I begin with a brief review about who bears responsibility for initiating and implementing policy.

## Whose business is it?

In recent decades, the development debate been polarised into two broad camps, referred to as the 'top-down' approach and the 'bottom-up' approach. These are also associated with the relative role of the market and the role of state or public interventions. The 'top-down' approach tends to emphasise the role of the state in leading the fight against hunger and poverty, while the 'bottom-up' school of thought prioritises the role of grassroots organisations and their ability to eliminate poverty and hunger. This debate has been unproductive and, as recognised by Chambers (1994), effective strategies to promote material improvements for the poor necessarily require changes from both directions. It is not very useful either to conclude that massive transformations in global capitalism are required if hunger is to disappear. While this may be true, and there is some merit in imagining such changes (imagining a better world is an essential prerequisite to establishing it), perhaps it is more useful to suggest changes which have a good chance of being effected in the rest of this decade than to have visions of a world turned upside down. The analysis now turns to considering some changes at every level which would have positive implications for hunger reduction. Knowing what changes would help is the relatively easy part. Assessing how to make the changes occur is more problematic and is considered last.

## Global changes

At the global level, changes in trade, aid and debt are most crucial. Changes in international trading could facilitate increased earnings in the South. The Uruguay Round will have a mixed impact (see Chapter 3) – it ignored many issues of concern to Southern nations. One of the most serious issues is the depression in world markets for primary products, the source of income for some thirty countries in Africa and eighteen in Latin America. 'Between 1980 and 1993, prices for non-oil primary commodities fell by more than half relative to prices for manufactured goods. The estimated annual loss to developing countries over this period was around $100 billion: more than twice the total flow of aid in 1990' (Watkins, 1995, 130). Stabilising commodity prices allied to efforts to diversify economies in the most vulnerable Southern nations is essential to changing this dependence and to enhancing these nations' international economic viability.

Among the most difficult international trading regulations to reform are those which challenge important interests in the North, for example manufacturing and agricultural interests. Despite the rhetoric of free trade, a range of legislative options is available to dominant Northern interests and Northern governments to protect the industrial, agricultural and service sectors of their economies against competition in the South. Protectionism takes a multitude of forms but its end result is that it tends to reduce the market opportunities of Southern producers. Reductions in these would allow Southern producers to generate more income and reduce poverty. The debate about the Multi-Fibre

Arrangement is very significant for some of the poorest Third World countries. Some of the implications of the international agricultural regime for Southern food security have been outlined earlier. Changes in this sphere of the international economy are overdue. Among the most urgently required changes are a reduction in agricultural subsidies in the North; a revision of agricultural policy which would promote environmental and ethical practices in the North and food security in the South simultaneously; and the implementation of innovative agricultural support policies in the South to enhance food security, rural employment and sustainability. New directions in agricultural policy are slowly being sought and fought for in both North and South as existing systems begin to expose their environmental and social externalities. The costs of energy-intensive agriculture are being exposed at the same time as increased productivity is revealed to have a complex relationship to reductions in hunger. Increases in food production are only part of the picture: increases in food production are sometimes correlated with increased landlessness and the *exacerbation* of hunger. In other words, extra food produced may be processed into less efficient foodstuffs associated with élite diets and therefore have no impact on hunger reduction.

While it is true to say that increasing the amount of food produced is not the answer to hunger alleviation, it would be foolish to deny that food production must be maintained, and the ramifications of drastic shortfalls are very serious because of the global nature of the contemporary food system (Box 8.1). Neither does a critique of modern farming methods necessarily reflect a Luddite mentality. New techniques in biotechnology and allied scientific advances in transport, storage, processing, etc., have potentially positive contributions to make to the elimination of hunger, but their potential remains largely unrealised. Current research in food production and technology is not *designed* to reduce the problem of hunger. The TNCs which dominate global research in this area respond to different agendas, but these could change given the political will by international agencies and customers.

The chronic debt crisis discussed earlier in this text is international in nature and it is a major factor which continues to deprive governments in the South of the funds to invest in infrastructure to improve income generation. Since 1987, the South has been transferring more capital, in debt repayments, to the North than has been transferred from the North to the South. International negotiations to resolve these debts without the poor in the South having to pay the price would be an important element of any international effort to alleviate world hunger. International negotiations about debt reductions have been too limited so far and flawed because they are tied to SAPs.

Box 8.1

## Trends in world food production

Trends in world per capita cereal production for the period 1951–93 are shown in Figure 8.1 (Dyson, 1996, 59). Steady increases in cereal production occurred, from 300 kg per capita in 1951 to over 350 kg in the 1990s. This figure confirms that, despite world population doubling between these dates, total cereal production per capita has increased markedly in the last thirty years. Most of this increase is due to increases in yields, not increases in area harvested; in fact the per capita area cultivated declined from about 0.235 to 0.127 hectares. Although there is a gradual increase in annual cereal production, occasionally there are sharp fluctuations in annual harvests. Climatic changes are often important, and if there is a drought in one of the world's major producing regions – North America, Europe, China or Russia – then the implications for the rest of the world are serious. The impact of the Chinese harvest failures between 1958 and the early 1960s is obvious on the graph of world totals. Another crisis of production on a global scale occurred in

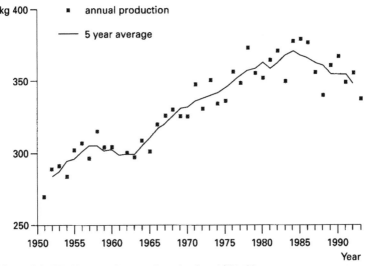

**Figure 8.1** World per capita cereal production, 1951–93

*Source*: Dyson (1996)

*Note*: Averages for 1952 and 1992 are calculated from data for 1951–53 and 1991–93, respectively (after Dyson, 1996).

Box 8.1 (*continued*)

1972–74, and its impact illustrates the importance of international factors to food security.

The crisis in 1972–74 was precipitated by several factors. Because cereal stocks were high in the early 1970s, farmers in the USA and Canada were encouraged to reduce the area planted to wheat; then in 1972 several major producer regions suffered a drought. Countries which were usually self-sufficient in grains became importers. The former Soviet Union, for example, purchased large quantities of wheat on the international market, and the price of wheat doubled between 1972 and 1974. It was not until 1977 that prices fell to their pre-crisis level. This crisis, of cereal production and price rises, was marked by increased mortality rates in Ethiopia, the Sahel, India, China and Bangladesh.

Like all aggregate figures, these figures of global cereal production must be interpreted carefully; when disaggregated by region other trends emerge. In sub-Saharan Africa yields per capita have been declining. Droughts in 1983, 1984 and 1992 have contributed to the problem of production but more serious is a combination of international and national factors, outlined elsewhere in this text. South Asia and the Far East have experienced increases in production per capita, and gains were especially impressive in China in the early 1980s.

Fifty per cent of the world's cereals are produced in North America, Latin America, the former Soviet Union and Europe. Declines in per capita production since 1984 are linked to changes in agricultural policy in North America and the EU particularly. As prices were falling and stocks were increasing, governments in both regions reduced incentives to farmers to grow cereals.

International aid, multinational or bilateral, has had a chequered history and although open to abuse it is still a potentially useful mechanism whereby capital and opportunities can be directed to the poor in the South. When carefully invested the livelihoods of the poor can be enhanced. The importance of aid in the budgets of Southern governments is varied but it remains particularly vital for some of the very poorest countries; in sub-Saharan Africa aid flows represent on average 13 per cent of national income. The quantity of aid from bilateral and multilateral sources has declined in recent years as the recession bit in the North. The UN's target is that Northern governments should transfer 0.7 per cent of their GNP (gross national product) in the form of aid to the

South. The list below (Watkins, 1995, 194) shows how distant this is from reality for most of the donor countries. The five highest and lowest contributors in 1993, as a percentage of GNP, have been selected from the developed economies.

| | |
|---|---|
| Denmark | 1.03 |
| Norway | 1.01 |
| Sweden | 0.98 |
| Netherlands | 0.82 |
| France | 0.63 |
| Japan | 0.26 |
| New Zealand | 0.25 |
| Spain | 0.25 |
| Ireland | 0.20 |
| United States | 0.15 |
| Average | 0.30 |

As well as the quantity of aid transferred, the quality of aid varies. Its relevance to poverty reduction or entitlement provision is often limited because other objectives, of donor or recipient government, are prioritised. The case of the Pergau Dam and British aid to Malaysia illustrates the point.

Aid is often tied, which means that it is granted only if the recipient country agrees to certain conditions. These conditions frequently promote interests in the donor country and Southern élites rather than the interests of the poor in the recipient country. The Pergau affair in Britain illustrates some of the potential problems. The British government undertook to provide £234 million finance over fourteen years to the Malaysian government to construct a dam on the Pergau River. Despite an assessment by development experts that the project represented 'a very bad buy' for the Malaysian government and British taxpayers, the project was funded. The project would provide relatively expensive electricity in Malaysia and suffered from all the social and environmental problems associated with such massive schemes. Yet it proceeded. Two issues help to explain why this project was funded: because it was negotiated in association with an arms purchase by Malaysia from Britain and because the construction contract was with British industrial interests.

This is not a problem unique to Britain, nor is it an exceptional case. Despite resolutions to reduce the level of tied aid it continues as donor countries compete for 'development business', which can end up promoting more jobs in the donor country than in the recipient country. Unfortunately too, two potential useful tied components, that aid should be conditional on the observance of human rights and on sustainability, are often ignored or circumvented in enthusiastic negotiations for industrial contracts. Instituting mechanisms to ensure that aid is targeted at the poor to enhance their entitlements would help to reduce hunger and perhaps starve some of the less environmentally and socially sound projects instead.

International financial institutions govern many of the rules of the international game, and reform of their policy and practice is essential if the rights of the poor and their prospects are to improve. Three of the most important are the WB, the IMF and the World Trade Organization (WTO), which has replaced the GATT. The World Bank is the main source of development funds for Southern governments. This grants it inordinate power, which it could use more effectively to enhance the prospects of the poor. In recent years, the Bank's policies have been modified to reflect increased concern about human rights, gender and environmental impacts of the projects it funds, but the implementation of these policies still tends to be determined by other interests. Watkins (1995, 203) concludes that while the Bank is better than many donor governments in assessing the impact of its projects, it may still ignore some serious problems – most recently in its funding ($770 million) of the large-scale Arun Dam in an ecologically unique area of Nepal inhabited by ethnic minorities, who will have to be relocated if the dam is built. Large-scale programmes have a poor record; they often exacerbate poverty and landlessness and always have dubious and occasionally disastrous environmental impacts. Discussion in Chapter 3 reviewed one element of IMF policy, its SAP programmes. The lending policies of the IMF and the conditions it requires of recipients must be changed to reduce their impact on poverty in the South. Decision-making in both these institutions reflects the concerns of the governments of the North and must be modified to reflect the needs of the recipient countries.

Finally, at the international level there are signs that attention, if not resources, is at least being given to the issues which form the focus of this text. It is easy to be cynical about international conferences devoted to global problems; after all, they provide photo opportunities for politicians and film stars and career advancement opportunities for academics and policy-makers. Participants at international conferences concerned about hunger and poverty are seldom poor or hungry. Such conferences may, nevertheless, be useful in the fight against hunger. They gather together some individuals and institutions genuinely devoted to the elimination of want in the world and, in addition, resolutions are made, financial and policy decisions are debated and concluded, targets are established, and evaluations of progress are made public. Related to this last point, maybe the most important point is that representatives from governments, and the myriad institutions and groups, are required to justify their policies and programmes and, often in the midst of media attention, explain their failures. For some representatives this is a novel experience (Box 8.2).

Box 8.2

## World Summit for Children 1990, United Nations in New York

The declaration and commitments:

- ratify, implement and monitor the Convention of the Rights of the Child;
- work for national and international action to enhance children's and mothers' health, universal clean water, and access to sanitation;
- eradicate hunger, malnutrition and famine;
- strengthen the role of women, maternal health, family size, child-spacing, breastfeeding, and safe motherhood;
- support the role of the family;
- reduce illiteracy and provide educational opportunities for all children;
- protect children in especially difficult circumstances;
- protect children in situations of armed conflict;
- preserve the environment;
- work for a global attack on poverty through national action and international cooperation.

Goals and targets for the year 2000:

- reduction of the death rate for children under five years of age to one-third of present levels (or below 70/1,000 live births, whichever is less);
- reduction of maternal mortality by one-half;
- reduction of severe and moderate malnutrition in children under five years of age by one-half;
- universal access to safe drinking water and sanitation;
- universal access to basic education and completion of primary education by at least 80 per cent of the population;
- reduction of the adult literacy rate by at least one-half, with emphasis on female literacy;
- improved protection of children in especially difficult circumstances.

While one of these goals is aimed directly at nutrition, all are relevant for its improvement. The seven supporting goals for nutrition are:

Box 8.2 (*continued*)

- reduction in the rate of low birthweight (less than 2.5 kg) to less than 10 per cent of all births;
- reduction of the incidence of iron deficiency anaemia in women by one-third of 1990 levels;
- virtual elimination of iodine deficiency disorders;
- virtual elimination of vitamin A deficiency and its consequences, including blindness;
- empowerment of all women to breastfeed exclusively for four to six months after their child is born, and to continue breastfeeding, with complementary food, well into the second year;
- institutionalisation of growth promotion and its regular monitoring in all countries by the end of the 1990s;
- dissemination of knowledge and supporting services to increase food production and thereby ensure household food security.

Having agreed a declaration and established targets, all participating countries were required to prepare National Programmes of Action, to organise regional follow-up meetings and to monitor their progress.

Basic socio-economic indices are presented in the Human Development Report prepared by the United Nations Development Programme. Statistics for selected countries from the 1995 report are given in Table 2.13.

*Source*: Jonsson and Zerfas (1994).

### International Conference on Nutrition, Rome, 1992

A total of 1,387 persons, including delegates from 159 nations and the European Union, 144 regional and national non-governmental organisations, sixteen UN organisations, and eleven intergovernmental organisations (IGOs) participated in the conference, which culminated in the adoption of a World Declaration and Plan of Action for Nutrition. Nine broad themes and a set of general goals were established and have since, to varying degrees, been woven into policy and action by the participating countries, with technical and organisational assistance from the various UN agencies and IGOs.

Box 8.2 (*continued*)

Themes:

- incorporating nutritional objectives, considerations and components into development policies and programmes;
- improving household food security;
- protecting consumers through improved food quality and safety;
- preventing and managing infectious diseases;
- promoting breastfeeding;
- caring for the socio-economically deprived and nutritionally vulnerable;
- preventing and controlling specific micronutrient deficiencies;
- promoting appropriate diets and healthy lifestyles;
- assessing, analysing and monitoring nutritional situations.

Goals:

As a basis for the Plan of Action for Nutrition and guidance for formulation of national plans of action, including the development of measurable goals and objectives within time frames, it was pledged to make all efforts to eliminate before the end of this decade:

- famine and famine-related deaths;
- starvation and nutrition deficiency diseases in communities affected by natural and man-made disasters;
- iodine and vitamin A deficiencies.

Participants also pledged to reduce substantially within the decade:

- starvation and widespread chronic hunger;
- undernutrition, especially among children, women and the aged;
- other important micronutrient deficiencies, including iron;
- diet-related communicable and non-communicable diseases;
- social and other impediments to optimal breastfeeding;
- inadequate sanitation and poor hygiene, including unsafe drinking water.

All these are very worthy and the conference provided an important

Box 8.2 (*continued*)

---

forum for the participants to make contacts and extend their knowledge. However, the concerns noted by Messer are pertinent:

> ... the challenge remains how to get from declarations and plans of action to improved individual nutrition in households. For this, the lines between global and national policy on the one hand, and individual behaviors in households on the other, must be connected, via health and nutritional providers and government and non-government liaisons, who can form the channels to listen to community problems and demands and to involve community members in monitoring and controlling their own nutritional situations.
>
> (Messer, 1994, 83)

*Source*: Messer (1994).

---

The role of the media in presenting the problem of world hunger has been only briefly mentioned in this text (Chapter 1). They have not always been helpful in contextualising the hunger and poverty that they portray. There is no doubt, however, that contributions from an engaged and concerned media can be effective in highlighting political neglect, corruption and injustices. In emergency situations, too, they are effective in raising concern and consequently funds for relief. International and national media have contrasting but vital roles to play in promoting more informed debates and political accountability.

There is genuine concern about the way the media and indeed NGOs represent the people who suffer from hunger and poverty. This issue goes to the centre of debates in development about who has the power and influence to speak for, or represent, people. The danger is that the poor are often represented as passive victims of their poverty and as a homogeneous mass. I hope this text has not been guilty of that distortion. The reality in the South is that people's entitlements are very diverse and their strategies for surviving remarkable. There is evidence across the South of people challenging the circumstances which explain their hunger and poverty. The media should represent more of these activities.

## The role of national governments

Policy and practice at the national level can address hunger effectively given the political will; some examples have been discussed earlier. The diverse character of governments and their diverse legacies in the realm of ideologies and social relations mean general prescriptions are difficult. The debate about how the state should or could intervene to promote development is still contentious, although there are, at present, real practical limitations on how governments might act, because of external circumstances. Whatever the ideology of the state, whatever shade of capitalist or socialist or hybrid, their resources in many cases are declining and so, therefore, are many of their options. Both the optimism and the resources that underpinned the 'developmental state' have declined. So while it was possible to debate the relevance and efficacy of state-led programmes of poverty alleviation in the past, it is clear that in the 1990s their role has been modified and reduced in most cases. The role of the market, rather than the state, in provisioning the poor has been emphasised by the international financial institutions. However, in most Southern contexts the state remains important for our analysis of hunger reduction, not just in famine or crisis situations, where it may be vital, but in the relief of malnutrition by devising policies to mediate globalisation processes and to influence changes in society which protect the most vulnerable.

Chapter 4 discussed how, in many countries, national agricultural policy has traditionally prioritised increases in production and how this has tended to advantage the rural population with access to land or capital. Mention was also made of how agricultural changes rely on high energy inputs and are associated with unsustainable systems of production. In the South and North, more effort must be spent researching and implementing farming which heightens equity and is environmentally sustainable. Rhetoric from governments is impressive in this area, but in practice few are addressing these issues seriously, because vested interests resist necessary changes; market-driven reforms are preferred to socially or environmentally motivated ones.

Whether rural or urban, the vulnerable are a very heterogeneous group and national policies must be designed to recognise the diversity of their entitlements. To the long-established marginal groups in the rural areas (landless, peasant producers, female-headed households, regional or ethnic minorities) and those long visible in urban areas (informal workers, many of them new migrants who live in *barrios, favellas*, etc.) must be added new groups of urban dwellers and workers in the formal sector who have suffered drastic declines in their livelihoods since the 1980s as SAPs, recession and inflation have wreaked havoc.

This grim context – escalating poverty and insecurity accompanied by a reduction in funds – helps to explain the emergence of the concept of the enabling state. This concept requires much less of the state than in the 1960s and 1970s; rather than effect grandiose interventions (transformation of the peasant consciousness for example, or massive resettlement programmes), the state and its bureaucracies now tend to be modest in their objectives. There are several functions where centralised efforts, 'top-down' interventions, work best and should be the objectives of all states. Only states can resource and implement basic infrastructural provision; transport, health, education, energy, water and sanitation provision are best financed and planned at the national level. The influence of all these has direct and indirect implications for the poor and their health and nutrition. States are also best equipped to deliver standard inputs in standard environments; child nutrition and immunisation programmes or nationwide nutrition education programmes are good examples. The state should provide or subsidise basic technology which is cheap, easy to maintain and appropriate. This technology may be in any area of activity: health, education, agriculture, industry, housing, etc.

Chambers (1994, 118–21) identifies another vital role for government which is pertinent to our analysis – the provision of a variety of safety nets for the vulnerable in rural and urban areas in the event of entitlement collapse. This includes public works programmes to provide wages, and early interventions to keep food prices down and incomes up during bad times so that the poorest do not have to sell their assets. Famine early-warning systems and the distribution of emergency food aid are certainly areas where the state can be most effective.

There is another area where states are usually more competent than others – that is the maintenance of peace. Unfortunately, there are many examples of states which are unable or unwilling to maintain peace but in most cases, rightly or wrongly, they still monopolise access to power; that is reflected in their representation at the UN, their access to negotiations with international financial institutions, their control of the military and police force, etc. As Chapter 7 illustrated, war and civil unrest are still a major cause of hunger and deprivation. International and domestic influences must combine to reduce military expenditures and redirect these funds to positive, humanitarian investments. Identifying what should be done is relatively straightforward but effecting the changes suggested is considerably more difficult. It would be foolish to pretend that there is an answer. However, some potentially important changes are occurring in the North and the South, and these are considered next.

## Empowerment approaches

> ... the explosion of community-based, participatory, grassroots action in most Third World countries over the last three decades is most encouraging: everywhere one looks, people are organizing to fight erosion, increase production and incomes, create safety nets, supply credit and other inputs, improve mothers' and children's health etc.
>
> (Uvin, 1994, 927)

The promotion of education in the South does not transform institutions and inequities, but it may help to transform the attitudes and actions of the poor and can cause them to insist on changes in institutions and government which increase their command over food. Even slight increases in education, broadly conceived, can promote increases in self-esteem, which can encourage people to challenge society's givens and to act more effectively; these changes are associated with the concept of empowerment, which has become popular during the 1980s and 1990s in development theory and practice. This concept was introduced in Chapter 5 but is considered in more detail here.

The concept of empowerment is particularly associated with grassroots organisations for social justice and sustainable development. The concept is potentially very radical because, if realised, it enables the poor to understand the causes of their poverty and to challenge entrenched interests which help perpetuate it. Empowerment approaches to poverty alleviation require the poor to be not only effective participants but also initiators of programmes designed to promote their health and well-being. Certainly, throughout the South, there is evidence that this is happening and that empowerment has material manifestations, increasing the poor's access to the resources that secure entitlements to a livelihood, land, education, credit facilities, skills and knowledge. All effective mobilisations of people to change their circumstances are based on some degree of empowerment. The current popularity of the concept reflects the realisation that social change must be to some extent an outcome of people's will to challenge and transform their own conditions, that people must be agents, not objects, of social programmes. As the concept of empowerment becomes popular with everyone from the World Bank to authoritarian governments, the danger is that it may become an empty rhetorical concept, used by all because it means little or nothing at all.

## The role of non-government organisations (NGOs)

Empowerment, in theory, is the epitome of people-centred development, where real participation is enjoyed by the people involved. Enthusiasm for

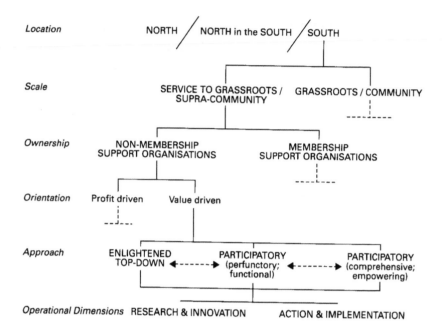

**Figure 8.2** Diversity of NGO types
*Source*: Farrington and Bebbington (1993, 21)

empowerment approaches has been inherently associated with the prominence of NGOs in development strategies in the 1980s and 1990s. NGOs come
in such a myriad of shapes and sizes that making generalisations about their
role or effectiveness is problematic. There are well-funded international
NGOs, staffed by professionals, national NGOs with diverse agendas, and
locally based NGOs, with political vision and commitment and some with
limited funds and objectives. Some important differences are shown in
Figure 8.2.

NGOs may be very radical, effect real political change and act as a
springboard for activists, but they are not necessarily so. What is indisputable
is that they are becoming more significant in development relief and aid. The
Organisation for Economic Co-operation and Development estimates that the
industrialised countries transferred $5.6 billion through NGOs to developing
countries in 1993 (Watkins, 1995, 207). Within the South they are expanding
too. The WB estimates that NGOs in India handle 25 per cent of all foreign

aid. To generalise about 'NGOs' is difficult because of the range in activities and ambitions, political philosophies and economic clout. Some international and national NGOs are important lobbyists (Box 8.3). Their activities at the International Conference on Nutrition (ICN, held in Rome, 1992, see Box 8.2) is typical of the way they may influence affairs. They used that occasion to form the Global Food and Nutrition Alliance and described their role as helping to identify the most urgent nutritional needs, and mobilise households, communities, and all forms of organisation as well as the mass media to help governments and individuals to implement the plans launched at the conference.

**Box 8.3**

---

**Brazil's campaign against hunger and misery, and for life**

Brazil: some statistics

| | |
|---|---|
| Population, 1992 | 153.8 million |
| Percentage urban, 1992 | 76 |
| Real per capita GDP | $5,240 |
| Percentage population in poverty | urban 38 per cent; rural 66 per cent |
| Daily calories per day per capita | 2,824 |
| Infant mortality rate | 58 per 1,000 live births |
| Access to safe water | urban 95 per cent; rural 61 per cent |
| Access to sanitation | urban 84 per cent; rural 32 per cent |

These statistics must be interpreted with care, because Brazil is a very polarised society; some of the regional contrasts were discussed in Box 2.3. The problem of street children in Brazilian cities has received publicity recently in the Western press, as has the problem of mugging and street crime against tourists. These are symptoms of the structural inequalities which have characterised Brazil since the colonial era. As many as 20 per cent of the population are estimated to be living in poverty, unable to meet their basic needs. The Campaign Against Hunger was formed in 1993 and is a broad-based alliance designed to raise awareness about hunger and to mobilise action to address it within Brazil. It encourages practical action at the local level and wider action to mobilise popular support for changes in government policy to alleviate the structural causes of poverty in the country.

Box 8.3 (*continued*)

There are now over 3,000 local committees and some 30,000 volunteer workers. The activities of these groups have ranged from growing food on neglected land to accepting food or money donations from local trade unions, churches and schools and building houses for the poor. Response has been positive and 3.5 million people have donated cash and 21 million food or clothes. In 1994 the campaign concentrated on unemployment and low wages, while in 1995 it publicised the plight of the rural poor, especially the landless, and issues of agrarian reform. Since its inception the campaign has ensured that awkward questions are raised of government policy and that officials are made accountable.

The most important achievement may be that cited by one of its founders:

[The campaign] made citizens and society begin to take responsibility for a problem that up until now had been treated only as a question to be resolved by those who are hungry and by those who govern. This change in perception, that understanding hunger and misery are items on society's agenda, is a major development in the public life of a country that, when it looks in the mirror, sees itself as apathetic, unethical, indifferent, selfish, and cunning.

(Herbert de Souza, quoted in Watkins, 1995)

*Source*: Based on Watkins (1995, 212–13).

Since independence from Pakistan in 1971, Bangladesh has faced some of the most pressing problems of poverty, famine and food insecurity. The World Bank judges it one of the five poorest countries in the world and the poorest in South Asia. An expert recently suggested that there 'are probably more NGOs in Bangladesh than in any other country of the same size in the world' (Lewis, quoted in Farrington and Bebbington, 1993, 50); these reasons explain why it is worth more analysis. The economy is typical of a poor country. It depends on only two crops, rice and jute, for its export earnings and relies on migrant remittances and large annual foreign aid subsidies to support the resident population. Over 80 per cent of the population of 113 million is rural; approximately 50 per cent of these people are landless and survive by recourse to a 'seasonal portfolio' of income-earning activities. The country is prone to floods and cyclones, so numerous NGOs have emerged to organise emergency relief operations but it often happens that these NGOs survive the emergency and begin to address longer-term development projects.

Beneficiaries of government agricultural extension services and international agricultural agencies have been rural people with access to land. These efforts have succeeded in increasing rice and wheat production, so that Bangladesh is now almost self-sufficient in cereals. However, the problems which remain exemplify the flaw in such approaches: the rural poor *without* access to land have very limited entitlements and cannot therefore afford to buy the extra food which is produced. Poverty among the rural landless, especially women, is extreme; it is among these people that some NGOs have concentrated their efforts.

The two largest NGOs in Bangladesh are the Bangladesh Rural Advancement Committee (BRAC) and Proshika. Both employ the 'target group' approach and in each case have concentrated on the rural poor, especially women. Their activities have two objectives: to increase income-earning opportunities and to empower the poor to challenge existing inequalities through education, organisation and mobilisation. Active since 1972, BRAC is national in scope and its objectives are poverty alleviation and the empowerment of the very poor, especially women. Both promote home-based income-generating activities such as cattle- and poultry-rearing, food-processing, social forestry, apiculture and rural handicrafts. Providing credit facilities to people who would otherwise be denied credit has been a crucial element of these initiatives. The strength of these approaches is that the poor do experience empowerment and some increases in income, and for numerous individuals these are very real benefits.

Proshika helped Basana Rani Bhaumik in Bangladesh, for example. She is a resident of Jamaya village in Manikgonj District and earns Tk2,000 per month by vaccinating chicken and cattle. She charges Tk0.5 per chicken vaccination and Tk1–2 for cattle. Basana gets the vaccines from Proshika. In addition, she has a contract with five surrounding villages, which give her 18 maunds of paddy rice at harvest time in lieu of payment for her services each season. Creation of such demand via the extension service is a unique innovation by Proshika. Similarly, heifer-rearing has made Rawsan Ara Kaunipara an impressive example of a landless woman improving her economic and social position. She has managed to earn Tk50,000 through selling the milk and the calf after three years (Farrington and Bebbington, 1993, 62).

However, several weaknesses are apparent. In the long term, without changes in government policy, NGOs' interventions may be unsustainable, and even the largest NGO in Bangladesh reaches a maximum of 20 per cent of the rural poor. More significantly, the activities of these NGOs have not altered local or national power structures: land ownership remains both concentrated and inequitable and patronage still prevails in rural and urban politics. These political realities have meant that some NGOs' projects have been obstructed or restricted in their impact.

Some NGOs have limited aspirations and their role as agents of empowerment may be minimal. Where they are effective the problem of how to expand their impact (the process of scaling up) without eroding the factors which explain their effectiveness has generated interest. Uvin (1994) examines how a variety of grassroots organisations that fight hunger have tried 'scaling up' and identifies four ways that organisations might increase their impact.

## Conclusion: concerted actions

Evaluating the distinct roles of actors/agencies at the international, national, grassroots or NGO level in eliminating hunger tends to ignore the fact that successful intervention requires their combined actions. The great variety of participants in international conferences in recent years reflects the fact that all levels must be involved and that diverse strategies can be employed to address the variety of issues, directly and indirectly, related to hunger in its myriad forms.

It is naive to assume too great a role for education in the process of hunger elimination, but equally it is cynical to ignore its potential if we understand the term in its widest sense and recognise a variety of places where it makes a positive contribution. Education and associated action in the North has a role to play and can change people's behaviour as consumers and voters. Understanding the role of global transactions and national policies in the creation of world hunger may help initiate changes to those policies in the North. A great variety of institutions in the North are lobbying for change and to promote a greater appreciation of the issues. Their role is important. The Hunger Program at Brown University is a good example, and the work of Oxfam is another.

It is useful to identify three aspects which help to explain social action to eliminate hunger. The poor and hungry fight to change the circumstances which increase or perpetuate their poverty; their motivation needs no explanation and that struggle is basic, often effective and always essential. Their challenges to 'the way things are' are a vital ingredient of any reduction in hunger. However, there are also millions in affluent circumstances, in the South and the North, who are concerned to change things. Their motivation is perhaps more difficult to divine, but most are motivated by a sense of justice which may be rooted in humanism or religion: all major world religions declare that the wealthy have a responsibility to the poor in their communities. The third aspect is that poverty relief usually reflects an element of self-preservation – if the poor become too many or too angry then stability is threatened and the prosperous suffer too. Also, given that hunger and disease are correlated, the threat from the poor and hungry takes a number of forms, from contagion to crime – both cause alarm among the wealthy. Even if, as they do, the well-off live in distinct areas, crime and disease are effective saboteurs of boundaries.

Common themes emerge at every level of analysis. The importance of inherited circumstances is clear. At the international level, contrasts in the power to influence Bretton Woods institutions and therefore the shape and direction of the world political economy, allied to economic circumstances inherited from the colonial era, mean that countries in the South have limited entitlements. At the national level, inherited social relations, e.g., inherited property relations, often mean that benefits from technological changes and expanded production accrue to a limited and already privileged minority. Regions often suffer the negative legacies of patterns of resource exploitation in the past, such as soil erosion and exhaustion from monoculture, as well as from limited or obsolete infrastructural investment. Individuals born into households with limited entitlements struggle to retain what few they have, often in challenging circumstances of social and economic change.

Ideologies embedded in social relations at every level explain contrasts in entitlements. Ideologies of the free market, exemplified by the SAPs, have been influential in structuring bilateral and multilateral lending policies at the international level since the 1980s. Western development ideologies since the 1950s, although different in emphasis, have prioritised economic expansion despite evidence that, in inequitable circumstances, this tends to exacerbate social and spatial differentiation. The propensity of policy-makers to assume that improvements in household incomes would benefit all its members leads to misguided policies that have disadvantaged women and children and their nutritional well-being. Patriarchal attitudes prevail at all levels and help to explain the negative impact of numerous policies on both food production and nutrition.

Of particular significance is the continued prevalence of ideologies in the North which understand poverty and hunger 'over there' as unconnected to conditions 'over here'. Because limited geographical and temporal frameworks are employed, simple apolitical interpretations abound. By theorising hunger as a problem of lack of entitlements, broadly conceived, the politics of food is exposed. This framework encourages us to understand hunger as a problem which exemplifies processes in the contemporary world economy where the local and global are connected through a complex chain of social processes. Granted the social character of hunger, we can more effectively evaluate policy interventions designed to reduce it at various scales. This framework is a useful introductory perspective. Equally important are analyses of how the politics of food has been transformed. From the decline of famine and malnutrition in eighteenth-century Europe to improvements in health and nutrition in some countries of the South in recent decades, evidence exists to illustrate how and why malnutrition is preventable. Some positive changes have been identified in this text: places and spaces within which people effectively challenge prevailing material inequalities and the ideologies which perpetuate them. We must hope that the positive changes identified expand and multiply, to coin a phrase.

## Key ideas

1  Changes at all levels, international to the household, are required to eliminate hunger. This includes changes in the way we understand the problem and the ideologies which perpetuate inequalities.
2  Some positive changes in the quality of life of the poor in the South have occurred in the last two decades.
3  Food security and sustainable development are inextricably linked.
4  Empowerment of disadvantaged groups is vital to the elimination of hunger.
5  Innovative policies by governments and NGOs are necessary to eliminate hunger.
6  Increases in food production will not solve the problem of world hunger. Increasing and securing people's entitlements will reduce hunger, and that requires policy changes at all levels of analysis.

# Review questions, references and further reading

## Chapter 1

### Review questions

1 Describe what is meant by the term 'entitlement' as detailed in Chapter 1.
2 What were the main killer diseases in nineteenth-century Britain and what factors help explain their prevalence in the past?
3 Why did mortality rates fall in Britain in the late nineteenth century? How did their decline vary socially?
4 What constitutes a famine? Describe with reference to a famine in the twentieth century.
5 Is there a 'world food problem' or a 'problem of world hunger'? Justify your answer with reference to contemporary circumstances.

### References and further reading

Arnold, D. (1988) *Famine: Social Crisis and Historical Change*, Oxford: Blackwell.

Bennett, J. and George, S. (1987) *The Hunger Machine*, Oxford: Blackwell.

Bonanno, A., Busch, L., Friedland, W., Goveia, L. and Mingione, E. (eds) (1994) *Columbus to Conagra: The Globalization of Agriculture and Food*, Kansas: University of Kansas Press.

Carey, J. (1987) *Eye-Witness to History*, Cambridge, Mass.: Harvard University Press.

Cook, I. (1994) 'New fruits of variety: symbolic production in the global food economy', in Bonanno *et al.* (eds) (1994) *op. cit.*, 232–48.

Crow, B. 'Understanding famine and hunger', in T. Allen and A. Thomas (eds) *Poverty and Development in the 1990s*, Oxford: Oxford University Press.

Daly, M. (1986) *The Famine in Ireland*, Dundalk: Dundalgan Press.

Devereux, S. (1993) *Theories of Famine*, London: Harvester Wheatsheaf.
Dreze, J. and Sen, A. K. (1989) *Hunger and Public Action*, Oxford: Clarendon.
—— (eds) (1990) *The Political Economy of Hunger*, vols 1–3, Oxford: Clarendon.
Lappe, F. M. and Collins, J. (1986) *World Hunger: Twelve Myths*, New York: Grove.
McManners, J. (1981) *Death and the Enlightenment*, Oxford: Clarendon.
Momsen, J. (1991) *Women and Development in the Third World*, London: Routledge.
Sen, A. K. (1981) *Poverty and Famines: An Essay on Entitlement and Deprivation*, Oxford: Clarendon.
Smith, F. B. (1979) *The People's Health, 1830–1910*, Croom Helm.
Warnock, J. W. (1987) *The Politics of Hunger*, London: Methuen.
Watts, M. J. and Bohle, H. J. (1993) 'The space of vulnerability: the causal structure of hunger and famine', *Progress in Human Geography* 17(1): 43–67.
Young, E. M. (1996a) 'World hunger: a framework for analysis', *Geography*, no. 351, vol. 81, Part 2, 97–110.
—— (1996b) 'Spaces for famine: a comparative geographical analysis of famine in Ireland and the Highlands in 1840s', *Transactions of the Institute of British Geographers* 21(4): 666–80.

## Chapter 2

### Review questions

1  What problems complicate the definition and measurement of hunger?
2  Describe the global incidence of hunger.
3  Evaluate some of the connections between disease and malnutrition and describe some of the diseases specifically associated with hunger.
4  What is 'hidden hunger' and why is it proving easier to address than under-nourishment?

### Further reading

Allen, T. and Thomas, H. (1992) *Poverty and Development in the 1990s*, Oxford: Oxford University Press/The Open University.
Arnold, D. (1988) *Famine: Social Crisis and Historical Change*, Oxford: Blackwell.
Bernstein, H., Crow, B. and Johnson, H. (1992) *Rural Livelihoods, Crises and Responses*, Oxford: Oxford University Press/The Open University.
Devereux, S. (1993) *Theories of Famine*, London: Harvester Wheatsheaf.
Dyson, T. (1996) *Population and Food: Global Trends and Prospects*, London: Routledge.
FAO/WHO (1992) *Nutrition and Development – A Global Report*, Rome: FAO/WHO.
Grigg, D. (1993) *The World Food Problem*, Oxford: Blackwell.
Jowett, A. J. (1987) 'Famine in the People's Republic of China', University of Glasgow, Department of Geography, Occasional Paper no. 21.
Smyke, P. (1991) *Women and Health*, London: Zed Books.
UNDP (United Nations Development Programme) (1995) *Human Development Report*, Oxford: Oxford University Press.

UNICEF (1994) *The State of the World's Children*, Oxford: Oxford University Press.

Uvin, P. (ed.) (1994) *The Hunger Report: 1993*, Brown University/Alan Shawn Feinstein World Hunger Program.

Warnock, J. W. (1987) *The Politics of Hunger*, London: Methuen.

Watkins, K. (1995) *The Oxfam Poverty Report*, Oxford: Oxfam.

# Chapter 3

## Review questions

1 Review some of the major changes occurring in the international food system and evaluate their implications for the problem of world hunger.

2 Contemporary agricultural production methods have been referred to as 'agri-cultural madness'. Do you think this term is accurate? Justify your answer with reference to current agricultural practice.

3 Identify some important historical legacies which continue to shape the contemporary world food system.

## References and further reading

Allen, T. and Thomas, A. (eds) (1992) *Poverty and Development in the 1990s*, Oxford: Oxford University Press.

Bernstein, H. B., Crow, B. and Johnson, H. (eds) (1992) *Rural Livelihoods: Crises and Responses*, Oxford: Oxford University Press.

Bernstein, H. *et al.* (1990) *The Food Question: Profits versus People?* London: Earthscan.

Bonanno, A., Busch, L., Friedland, W., Gouveia, L. and Mingione, E. (eds) (1994) *Columbus to Conagra: The Globalization of Agriculture and Food*, Kansas: University of Kansas Press.

Goodman, D. and Redclift, M. (1991) *Refashioning Nature*, London: Routledge.

Grigg, D. (1993) *The World Food Problem*, Oxford: Blackwell.

Heffernan, W. D. and Constance, D. H. (1994) 'Transnational corporations and the globalization of the food system', in Bonanno *et al.* (eds) (1994) *op. cit.*, 29–51.

Hobbelink, H. (1991) *Biotechnology and the Future of World Agriculture*, London: Zed Books.

Johnston, R. J. and Taylor, P. J. (eds) (1989) *World in Crisis*, Oxford: Blackwell.

Knox, P. and Agnew, J. (1994) *The Geography of the World Economy*, London: Edward Arnold.

McMichael, P. (1996) *Development and Social Change: A Global Perspective*, London: Pine Forge Press.

Madeley, J. (1992) *Trade and the Poor*, London: Intermediate Technology Publications Ltd.

Mies, M. and Shiva, V. (1993) *Ecofeminism*, London: Zed Books.

Riley, S. P. and Parfitt, T. W. (1994) 'Economic adjustment and democratization in Africa', in Walton and Seddon (eds) (1994) *op. cit.*, 135–70.

Shiva, V. (1991) *The Violence of the Green Revolution*, Penang: Third World Network.
Tansey, G. and Worsley, T. (1995) *The Food System: A Guide*, London: Earthscan.
Taylor, P. J. (1992) 'Understanding global inequalities: a world-systems approach', *Geography*, 77(1), 10–21.
Walton, J. and Seddon, D. (eds) (1994) *Free Markets and Food Riots: The Politics of Global Adjustment*, Oxford: Blackwell.
Wells, T. (1995) *The New Internationalist Food Book*, Oxford: New Internationalist Publications Ltd.
Williams, E. (1984) *From Columbus to Castro: The History of the Caribbean*, New York: Random House.

## Chapter 4

### Review questions

1 Evaluate the variety of factors which help shape any government's agricultural policies.
2 Assess the dangers to national food security which may be associated with a too limited export base.
3 With special reference to a country in the South compare and contrast the factors which influence the entitlements of its urban and rural populations.

### References and further reading

African National Congress (May 1994) Agricultural Policy Document, summarised in *African Farmer*, Jan–April 1995, 30–2.
Barraclough, S. (1991) *An End to Hunger*, London: Zed Books.
Bernstein, H. B., Crow, B. and Johnson, H. (eds) (1992) *Rural Livelihoods: Crises and Responses*, Oxford: Oxford University Press.
Brown, L. (1995) *Who Will Feed China?* London: Earthscan.
Bryant, E. (1994) 'Farming fish: nutritious ponds bolster agriculture', *African Farmer*, 42–3.
Christodoulou, D. (1990) *The Unpromised Land*, London: Zed Books.
Croll, E. (1994) *From Heaven to Earth: Images and Experiences of Development in China*, London: Routledge.
Crow, B. (1992) 'Rural livelihoods: actions from above', in Bernstein *et al.* (eds) (1992) *op. cit*, 251–73.
Drakakis-Smith, D. W. (1994) 'Food-systems and the poor in Harare under conditions of structural adjustment', *Geografiska Annaler*, 76 BI, 3–19.
El Ghonemy, M. R. (1990) *The Political Economy of Rural Poverty: The Case for Land Reform*, London: Routledge.
Elliott, J. (1994) *An Introduction to Sustainable Development: The Developing World*, London: Routledge.

Ghosh, J. and Bharadwaj, K. (1992) 'Poverty and employment in India', in Bernstein *et al.* (eds) (1992) *op. cit*, 139–64.

Hubbard, M. (1995) *Improving Food Security*, London: Intermediate Tech.

Leeming, F. (1993) *The Changing Geography of China*, Oxford: Blackwell.

Stonich, S. (1992) 'Struggling with Honduran poverty: the environmental consequences of natural resource-based development and rural transformation', *World Development*, 20(3), 385–99.

Thubron, C. (1988) *Behind the Wall: A Journey through China*, London: Penguin.

# Chapter 5

## Review questions

1 Is a gender-aware analysis helpful when analysing world hunger? Justify your answer with reference to case studies from the South.
2 Evaluate the relevance of family politics to an understanding of the problem of hunger.
3 Why do you think the importance of gender to analyses of hunger has been ignored for so long?
4 Suggest some changes in national policy which could help to reduce gender bias and thereby also help to reduce the incidence of hunger.

## References and further reading

Boserup, E. (1970) *Women's Role in Economic Development*, New York: St Martin's Press.

Kabeer, N. (1994) *Reversed Realities: Gender Hierarchies in Development Thought*, London: Verso.

Khatun, R. (1991) quoted in Alam, S., 'The control of girls,' *New Internationalist* (Nov), 15–17.

Mies, M. and Shiva, V. (1993) *Ecofeminism*, London: Zed Books.

Momsen, J. H. (1991) *Women and Development in the Third World*, London: Routledge.

Momsen, J. and Kinnaird, V. (1993) *Different Places, Different Voices: Gender and Development in Africa, Asia and Latin America*, London: Routledge.

Namakando, M. (1991) 'A song of the strong', *New Internationalist* (Nov), 10–12.

Nussbaum, M. and Glover, J. (1995) *Women, Culture and Development*, Oxford: Oxford University Press.

O'Connell, H. (ed.) (1994) *Women and the Family*, London, Zed Books.

Ostergaard, L. (1992) *Gender and Development: A Practical Guide*, London: Routledge.

Sen, G. and Grown, K. (1987) *Development, Crises, and Alternative Visions: Third World Women's Perspectives*, New York: Monthly Review Press.

Shiva, V. (ed.) (1994) *Close to Home*, London: Earthscan.

Sparr, P. (ed.) (1994) *Mortgaging Women's Lives*, London: Zed Books.

Synder, M. (1995) *Transforming Development: Women, Poverty and Politics*, London: Intermediate Technology Publications.
Tinker, I. (1990) *Persistent Inequalities: Women and World Development*, Oxford: Oxford University Press.
United Nations Development Programme (UNDP) (1995) *Human Development Report, Oxford: Oxford University Press.*
Visvanathan, N., Duggan, L., Nisonoff, L. and Wiegerma, N. (eds) (forthcoming) *The Women, Gender and Development Reader*, London: Zed Books.
Women's Feature Service (1993) *The Power to Change: Women in the Third World Redefine their Environment*, London: Zed Books.

## Chapter 6

## Review questions

1 With reference to a specific country describe and explain its patterns of malnutrition.
2 Why do minority populations suffer more from hunger than majority populations?
3 Evaluate the relationship between climate and hunger with specific reference to one region in the South.
4 What important factors influence the local food security system?

## References and further reading

Abeywardene, P. (1995) 'Food processing in Sri Lanka', in Appleton, H. (ed.) *Do It Herself: Women and Technical Innovation*, London: Intermediate Technology Publications.
Adams, A. (1993) 'Food security in Mali: exploring the role of the moral economy', *IDS Bulletin*, 24(4), 41–51.
Adams, W. M. (1990) *Green Development, Environment and Sustainability in the Third World*, London: Routledge.
Bernstein, H., Crow, B. and Johnson, H. (1992) *Rural Livelihoods: Crises and Responses*, Oxford: Oxford University Press.
Boserup, E. (1970) *Woman's Role in Economic Development*, New York: St Martin's Press.
Brown, J. L. and Pizer, H. F. (1987) *Living Hungry in America*, New York: Penguin.
Chambers, R. (1983) *Rural Development: Putting the Last First*, Harlow: Longman.
Croll, E. (1994) *From Heaven to Earth: Images and Experiences of Development in China*, London: Routledge.
Devereux, S. (1993) *Theories of Famine*, London: Harvester Wheatsheaf.
Elliott, J. (1994) *An Introduction to Sustainable Development: The Developing World*, London: Routledge.
Hubbard, M. (1995) *Improving Food Security*, London: Intermediate Tech.

Lipton, M. and Maxwell, S. (1992) *The New Poverty Agenda: An Overview*, Discussion Paper 306, University of Sussex: Institute of Development Studies.

Sen, G. (1992) 'Social needs and public accountability: the case of Kerala', in Wuyts *et al.* (eds) (1992) *op. cit.*, 253–77.

Sontheimer, S. (ed.) (1991) *Women and Environment: A Reader*, London: Earthscan.

Thubron, C. (1988) *Behind the Wall: A Journey through China*, London: Penguin.

UNDP (United Nations Development Programme) (1995) *Human Development Report*, Oxford: Oxford University Press.

Webb, P. and von Braun, J. (1994) *Famine and Food Security in Ethiopia. Lessons for Africa*, New York: Wiley.

Wuyts, M., Mackintosh, M. and Hewitt, T. (1992) *Development Policy and Public Action*, Oxford: Oxford University Press.

# Chapter 7

## Review questions

1  'The relationship between war and famine is direct, lethal and seemingly intractable' (Devereux, 1993, 163). Discuss this statement with reference to examples.

2  Discuss the international, national and local context of a major 'complex emergency'.

3  With reference to one contemporary 'complex emergency' describe the changes in entitlements for those caught in its scope.

## References and further reading

Black, R. and Robinson, V. (1993) *Geography and Refugees: Patterns and Processes of Change*, London: Belhaven.

de Waal, A. (1993) 'War and famine in Africa,' in New Approaches to Famine, *IDS Bulletin*, 24(4), 33–40.

Devereux, S. (1993) *Theories of Famine*, London: Harvester Wheatsheaf.

Duffield, M. (1994) 'The political economy of internal war: asset transfer, complex emergencies and international aid', in Macrae and Zwi (eds) (1994) *op. cit*, 222–32.

Forbes-Martin, S. (1992) *Refugee Women*, London, Zed Books.

Green, R. H. (1994) 'The course of the four horsemen: the costs of war and its aftermath in sub-Saharan Africa,' in Macrae and Zwi (eds) (1994) *op. cit.*, 37–49.

Keen, D. and Wilson, K. (1994) 'Engaging with violence: a reassessment of relief in wartime,' in Macrae and Zwi (eds) (1994) *op. cit.*, 209–21.

Macrae, J. and Zwi, A. (eds) (1994) *War and Hunger*, London: Zed Books.

Messer, E. (1994) 'Food wars: hunger as a weapon of war in 1993', in Uvin, P. (ed.) *The Hunger Report: 1993*, Brown University/Alan Shawn Feinstein World Hunger Program, 43–69.

Watkins, K. (1995) *The Oxfam Poverty Report*, Oxford: Oxfam.

## Chapter 8

### Review questions

1  At what level must change be initiated to relieve hunger effectively?
2  Evaluate the role of international NGOs in efforts to eliminate world hunger.
3  There is not a 'world food problem' but rather a 'world hunger problem'. Discuss the validity of this statement.
4  'Eliminating world hunger and promoting sustainable development are inextricably linked.' Debate this assertion.

### References and further reading

Chambers, R. (1994) *Challenging the Professionals: Frontiers for Rural Development*, London: Intermediate Technology Publications.
Clark, J. (1991) *Democratizing Development. The Role of Voluntary Organizations*. London: Earthscan.
Dyson, T. (1996) *Population and Food*, London: Routledge.
Farrington, J. and Bebbington, A. (eds) (1993) *Reluctant Partners? Non-Governmental Organizations, the State and Sustainable Agricultural Development*, London: Routledge.
Hobbelink, H. (1991) *Biotechnology and the Future of World Agriculture*, London: Zed Books.
Jackson, B. (1994) *Poverty and the Planet*, London: Penguin.
Jonsson, U. and Zerfas, A. (1993) 'After the World Summit for children: achieving the nutritional goals through national programs of action', in Uvin (ed.) (1994) *op. cit.*
Kirby, J., O'Keefe, P. and Timberlake, L. (1995) *The Earthscan Reader in Sustainable Development*, London: Earthscan.
Messer, E. (1994) 'The International Conference on Nutrition: historical perspectives and prospects', in Uvin, P. (1994) *op. cit.*, 71–85.
Reid, D. (1995) *Sustainable Development: An Introductory Guide*, London: Earthscan.
UNDP (United Nations Development Programme) (1995) *Human Development Report*, Oxford: Oxford University Press.
Uvin, P. (1994) *The Hunger Report: 1993*, Brown University/Alan Shawn Feinstein World Hunger Program.
Watkins, K. (1995) *The Oxfam Poverty Report*, Oxford: Oxfam.

# Index

Abeywardene, P. 127–9
acute hunger *see* famine
Adams, A. 118
Adams, W.M. 117
Afghanistan: consequences of conflict
142–3
Africa 3; fish farming 78; food riots
45; land management 116–17; land
ownership 66; low birthweights 25;
safe water supply 25; *see also*
specific countries
Africa, North: hunger statistics 27;
land ownership 67
Africa, sub-Saharan: aid 152; cereals
152; child malnutrition 105; conflict
144; entitlement 6; food wars 134;
hunger statistics 27–30; refugees
140; role of climate 115; women
agricultural workers 95, 96–7; *see
also* specific countries
Africa, West: hunger statistics 30
African National Congress:
agricultural policy 84–5
African-Americans: low birthweights
25
age: and hunger 34; low height for 25
Agnew, J. 41
agriculture: African National Congress
policy 84–5; cash crops 72–5; and

conflict 136–7; diffusion of
Western practices 54–9;
international policies 150; marginal
farming 115; national policies 159;
women workers 88–9, 91–8
aid, international: in conflict areas
134–5; and hunger statistics 26–7;
importance of 152–4; self-
preservation in 166–7; tied 153–4
aid agencies, international:
Afghanistan 143; and conflicts 144;
*see also* specific names
America, Central 27, 144; cash crops
72–4
America, North: cereals 151, 152;
entitlement 41; as integrated union
49
America, South 144; hunger statistics
27; *see also* Latin America; specific
countries
anaemia: in women 20–1
ancient world: food systems 37
animals: as food security; trees as
fodder for 129–30
aquaculture: Africa 78
Argentina: infant mortality rate 31;
wheat 38–40
arms purchase 141
Arnold, D. 3, 11, 12, 18

Asia, South: hunger statistics 27–30;
    land ownership 67; safe water
    supply 24–5; see also specific
    countries

babies see infants
Bangladesh: child malnutrition 105;
    empowerment approaches 164–6;
    hunger statistics 27; infant mortality
    rate 25; women's empowerment 107
Bangladesh Rural Advancement
    Committee (BRAC) 165
Barraclough, S. 64–6, 67–70, 74
Bebbington, A. 162, 164, 165
beef see meat
beneficiaries: of conflicts 135–6
Bernstein, H. 38–40, 66, 78, 113, 114
Black, R. 140
Bohle, H.J. 4
Bonanno, A. 36, 47, 52–3
Boserup, E. 88
Bostum, Rashid 143
Brazil 37–8; empowerment approaches
    163–4; hunger statistics 30; infant
    mortality rate 31–3; quantity of food
    64–6
bread: in nineteenth-century England 8
breadfruit trees 127–9
breastfeeding debate 23–4
Bretton Woods institutions (BWI) 43,
    70, 167
Britain: aid to Malaysia 153; cholera in
    nineteenth century 12–14;
    entitlement 6; food and drink in
    nineteenth century 8–9; fresh foods
    48; infant mortality rate 25;
    mortality decline 11–12
BSE scare 55, 58
Burkina Faso: cash crops 72, 74
Burundi 133–4, 140

Canada see America, North
Canton 120–1
capital, mobility of: and food systems
    48–9
Carey, J. 12–14
Cargill conglomerate 52–3
Caribbean: child malnutrition 105;

sugar production 37–8
Caribbean Basin Initiative (CBI) 73–4
cash crops 72–5
casualties: in conflicts 137
cattle sales: in conflict situations
    136–7; see also meat
causes of hunger 2–3; Great Irish
    Famine 8–10; proximate and
    structural 4, 10–11
cereal production: trends in 151–2; see
    also wheat
Chambers, R. 125–6, 149, 160
children: in Afghanistan 142–3;
    dietary deficiencies 19–20; female
    malnutrition rates 105; and
    malnourished mothers 21–2;
    marasmus and kwashiorkor 25–6;
    and mothers' status 89; positive
    improvements for 148; refugee 140;
    secondary malnutrition 23–4; World
    Summit for Children 20, 155–6
Chile: pastel de choclo (corn pie with
    chicken) recipe 132
China 3; cereals 151, 152;
    development strategies 71; gender
    bias 105; hunger statistics 27, 30;
    land ownership 67–70; regional
    contrasts 120–1
cholera 11; in nineteenth-century
    Manchester 12–14
chronic hunger see malnutrition
climate: role of 115–18
Collins, J. 3
colonisation, European: and
    entitlement 36–47, 66–70; and
    gender 95–6
common agricultural policy (CAP)
    75–8
common property resources (CPRs)
    115
conferences, international 155–8, 166
conflict: and hunger 4, 133–45; role of
    state 160; within households 102–3
Constance, D.H. 52–3, 59
Convention on the Rights of the Child
    24
Cook, I. 2
crime 167

Croll, E. 67–70, 120
crops: exotic 38; for export 72–5;
    failure 4, 7–10; staple 38–40; *see
    also* specific crops
Crow, B. 38–40, 66, 78, 113, 114
Cuba: infant mortality rates 31

Daly, M. 10
dams: construction 71–2; Pergau 153
De Souza, Herbert 164
De Waal, A. 138
debt crisis 42–3, 150
deficiencies, nutrient 19–21
deforestation 130
demography: imbalance as
    consequence of conflict 142–3;
    population growth in relation to
    hunger 2–4
desertification 116–17
developing countries: anaemia in
    women 21; assumptions about
    hunger in 3; debt crisis 42–3;
    diffusion of Western agricultural
    practices 54–9; empowerment
    approaches 161–6; entitlement
    concept 6, 41–2; food specialisation
    41; historical legacies 66–70;
    hunger statistics 27–30; local foods
    59, 60–1; role of trees 126–30
development programmes: dams 71–2,
    153; large-scale 154
development resources: diversion to
    arms purchase 141
development strategies: national
    70–83; top-down and bottom-up
    approaches 148–9; Western 167
Devereux, S. 3, 18, 115, 141
diarrhoea 23, 26
dietary deficiency malnutrition 19–21
disease: in conflict areas 138; public
    health reforms 12–15; and refugees
    140; relation to hunger 10–11, 18,
    167; and secondary malnutrition
    23–5; in semi-arid areas 116; and
    undernutrition 25, 26
displaced people 121; entitlement
    139–41
distibution: hunger 27–30

Drakakis-Smith, D.W. 79–83
drink, quality of: in nineteenth-century
    England 8–9; *see also* milk; water
drought: as cause of hunger 4; effect of
    117–18
Duffield, M. 135, 137
dysentery 26
Dyson, T. 28, 151

earnings, women's: in formal economy
    99; as primary income 102
economics: in conflict areas 136;
    international 149–52
economy: Afghanistan 142–3;
    informal, and female entitlements
    98–9
education 148; promotion of 161; role
    of 166
Egypt: household expenditure 103–4
Elliott, J. 71–2, 115
Ellisussen, Sofie 143
emergencies, complex 134
empowerment: approaches 161–6;
    women 106–9
England *see* Britain
Enlightenment: humanitarianism 15
entitlement concept: defined 4–7
environment: degraded or hazardous
    areas 115; desertification 116–17;
    and national agricultural policies
    159; and Western agricultural
    practices 54–5
Ethiopia 141
ethnic minorities 120–1; cleansing
    136
Europe, Western: cereals 151;
    colonisation and food systems
    36–47; decline of hunger 7–15;
    decline of mortality 11–14;
    entitlement 41
European Union: cereal production
    152; common agricultural policy
    75–8; dominance in agricultural
    trade 46; entitlement 6; as integrated
    market 49
expenditure, household: and gender
    103–4
export: crops for 72–5

famine: causes 4; in China 67; and
conflict 134; defined 18–19; early-
warning systems 160; Great Irish
Famine 7–10; mortality rates 17–18;
political context of statistics 26–7
Farrington, J. 162, 164, 165
financial institutions, international:
development funds 154; *see also*
specific names, e.g. World Bank
fish farming: Africa 78
fishing grounds: access to 115
floods: as cause of hunger 4
food: breastfeeding debate 23–4; in
nineteenth-century England 8–9; *see
also* specific foods
Food and Agricultural Organisation
(FAO) 19, 78; hunger statistics 27
food production: geography of 37–47;
relationship with hunger 150; trends
in 151–2
food riots: International Monetary
Fund 43–6; state fear of 15; in
Zambia 96
food system, global: fresh foods 48–9;
since 1970s 47–59; sixteenth to
twentieth centuries 37–47
food wars 134
Forbes-Martin, S. 140
formal sector: female earnings in 99
fresh foods: long-distance trade 48–9
fruit 126–9
fuelwood 129; depletion 130

gender: and entitlement 88–109; and
hunger 33–4
gender inequality: measures of 90–1
gender-related development index
(GDI) 91; changes in values 106
General Agreement on Tariffs and
Trade (GATT) 43; impact of 46–7
genocide, cultural 136
geography: world hunger 27–30
global changes, future 149–58
Global Food and Nutrition Alliance
163
globalisation: food systems 47–59
Glover, J. 107–9
governments *see* state

grain prices: in drought 117–18
Grameen Bank 107
grassroots organisations:
empowerment 161; role of 148–9;
women's empowerment 106
Green, R.H. 134
green revolution 42, 72
Grigg, D. 20
gross domestic product per capita
(GDP): entitlement 7
gross national product per capita
(GNP) 31; and HDI rankings 90;
and international aid targets 152–3
Grown, K. 88

health care: women's entitlement to
89; *see also* disease
health organisations: breastfeeding
debate 24; in conflicts 138
Heffernan, W.D. 52–3, 59
height for age: as indication of
undernutrition 25
hidden hunger 19–21
historical perspectives: decline of
hunger in Western Europe 7–15;
food access 36–59; national food
entitlement 66–70
Honduras: cash crops 72–4
households: entitlement within 5,
99–105, 167; food security in
121–30; peasant 68–9; poor and ultra-
poor 125–6
human development indices (HDI) 90;
and entitlement 7
*Human Development Report* 148
human rights 144
humanitarianism: in Enlightenment
15
hunger: alternative futures 147–68;
causes 2–3; and conflict 133–45;
contemporary extent 26–34;
framework for analysis 3–7; and
gender 88–109; hidden 19–21;
historical overview 7–15;
international perspectives 36–62;
national perspectives 64–86; sub-
national perspectives 111–31;
terminology 17–26

immunisation 148; role of state 160
imports: in conflicts 137–8
India 78; cholera 11; coconut curry
    recipe 146; dam construction 72;
    gender bias 105; hunger statistics
    27; infant mortality rate 25, 31; land
    ownership 66; mixed vegetable
    curry recipe 110; regional
    entitlements 112–15; safe water
    supply 25; Widows Conference
    108–9; women's empowerment
    107–8
Indonesia: household expenditure 104
industrial agriculture: diffusion 54–9
industrialisation 70–83
infant mortality rate (IMR) 31; Brazil
    31–3; female 105; improvements
    148; as indication of undernutrition
    25; Zimbabwe 44
infants: breastfeeding debate 23–4; and
    maternal anaemia 21
informal sector: female entitlements
    98–9; women's associations 107–8
information systems technology: role
    in global food system 52
infrastructural provision: dam
    construction 71–2, 153; role of state
    160
Innocenti Declaration 24
International Conference on Nutrition
    (1992) 156–8, 163
international conferences 154–8, 166
international level: entitlement 5, 6–7,
    40–2, 167; future changes 149–52;
    world hunger 36–62
International Monetary Fund (IMF):
    development funds 154; food riots
    43–6; and gender 91; industrial
    development 71
international relief see aid
iodine deficiency 20
Ireland: Great Famine 7–10; infant
    mortality rate 31
iron deficiency 20–1

jak trees 127–9
Japan 46; entitlement 41; infant
    mortality rate 25; land ownership 67

Johnson, H. 38–40, 66, 78, 113, 114
joint ventures: in global food system 53
Jordan: basal badawi (onions with
    meat, nuts and fruit) recipe 87
Jowett, A.J. 27
justice 166

Kabeer, N. 102
Keen, D. 136
Kenya: land ownership 94–5; roast
    chicken with peanut sauce recipe 63;
    women agricultural workers 94–5
Kerala, India 114–15
Knox, P. 41
Korea, North: floods 4
Korea, South: land ownership 67
kwashiorkor 25–6

labour: diversion in conflicts 141–2;
    women's 88–9
Ladurie, Le Roy 11
land management: Africa 116–17
land mines 137
land ownership: Argentina 39–40;
    Bangladesh 166; in conflict areas
    136–7; in developing countries
    66–70; Ireland 9–10; Kenya 94–5
Lappe, F.M. 3
Latin America: child malnutrition 105;
    food riots 45; land ownership 66, 67
life expectancy 148
Lipton, M. 115, 121
livestock: as food security 117–18;
    trees as source of fodder for 129–30
local food security 121–30; see also
    households
low birthweights: as indication of
    undernutrition 25
low height for age: as indication of
    undernutrition 25
low weight for height: as indication of
    undernutrition 25

Macrae, J. 133, 135, 137
Madagascar: food riots 44–5
malaria 21
Malaysia: aid to 153
Mali: food security strategies 118

malnutrition: Brazil 31–3; and conflict
134; defined 19–26; distribution
27–30; mortality rates 17–18;
political context of statistics 26–7;
secondary 23–5; and women 21–2
Malthusian interpretations of hunger 3
marasmus 25–6
Mauritius: beef rougaille recipe 35
Maxwell, S. 115, 121
McManners, J. 7, 10
meat: Argentina 39; in nineteenth-
century England 9
media: and conflicts 144; role of 158
medical technology: and decline in
mortality rates 12
men: diversion of labour in conflicts
141–2; see also gender
menarche, delayed: as indication of
undernutrition 25
Messer, E. 134, 155–8
Mexico: debt crisis 43; quantity of
food 64–6
micronutrient deficiencies 19–21
Middle East 144
Mies, M. 46, 98
migration: and conflict 136–7, 138; for
food 11; forced 138
military: arms purchase 141; as
beneficiaries of relief 136
milk: breastfeeding debate 23–4; in
nineteenth-century England 8–9
minorities: ethnic and religious 120–1;
ethnic cleansing 136; and hunger 34
modernisation theory of development
70–2
Momsen, J.H. 3, 89, 94, 98
mortality rates: from chronic and acute
hunger 17–18; decline in Western
Europe 7, 11–14; and disease
10–11; in eighteenth-century Europe
7; of Great Irish Famine 7–8;
women 89; see also infant mortality
rate
Mozambique: villagisation 138
Multi-Fibre Arrangement 149–50

Namakando, M. 96–7
Nasiba, Bibi 143

national level: entitlement 5, 40–2,
64–86, 167
natural causes of hunger 2
Near East: hunger statistics 27
Nepal: animal fodder 130; Arun Dam
154
newly industrialised countries (NICs):
entitlement 6–7
Niger: infant mortality rates 31
non-governmental organisations
(NGOs): relief operations 135; role
of 161–6; women's empowerment
106
Nussbaum, M. 107–9
nutrients deficiency: 19–21
nutrition education programmes 160
Nzioki, Elizabeth 94–5

O'Connell, H. 103–4
oil crises 42–3
Organisation for African Unity (OAU)
139
Organisation for Economic Co-
operation and Development (OECD)
162

Pakistan: hunger statistics 27
Parfitt, T.W. 44–5
peace, maintenance of: role of state 160
Philippines: land ownership 66, 67
pneumonia 23
politics of hunger 2–3, 4, 167–8; and
conflict 135–7; role of state 12,
14–15; Sahel desertification 117;
and statistics 26–7
poor see poverty
population see demography
potato: Great Irish Famine 7–10; as
staple food 38
poverty: Brazil 32–3; cycles of 3; in
households 125–6; in nineteenth-
century England 8–9;
representations of 158
power hierarchy: and food systems 41
profiteering: in conflict areas 135
Proshika 165
protectionism, trade: 149–50
public health reforms 12–15

public works programmes 160

Rabbani, Burhanuddin 143
Rashid, Ahmed 142
recipes: basal badawi 87; beef
    rougaille 35; coconut curry 146;
    mixed vegetable curry 110; pastel de
    choclo 132; roast chicken with
    peanut sauce 63; shrimp and chicken
    salad 1; sweet potato leaf stew 16
refugee camps 140
refugees 121; entitlement 139–41
regional level: entitlement 5, 111–21,
    167
relief see aid
religion 166
religious minorities 120–1
resettlement policies: in conflict areas
    138
revolutionary movements: state fear of
    15
Riley, S.P. 44–5
Rio Conference 117
Robinson, V. 140
rural areas: diversity in 78; entitlement
    6; empowerment approaches 165;
    India 112; lack of state funds 70–1
Russia: cereals 151, 152; cholera 11
Rwanda 133–4

Sahel: desertification 116–17
sanitation, safe: access to 24–5
Save the Children: Afghanistan 143
Scheper-Hughes, Nancy 33
scorched-earth tactics 137
seasonality: relevance of 115–16
Seddon, D. 42, 44–5
Self-Employed Women's Association
    of India (SEWA) 107–8
semi-arid areas 115–18
Sen, A.K. 4, 5
Sen, G. 88, 114–15
Shiva, V. 46, 98
Sierra Leone: sweet potato leaf stew
    recipe 16
Smith, F.B. 8–9
Smyke, P. 21
social action: to eliminate hunger

166–7
social dislocation: in conflicts 137; and
    disease 11; and famine 18; Ireland
    8–10
Sontheimer, S. 126
sourcing, global: and TNCs 53
South see developing countries
South Africa: agricultural policy 84–5
specialisation, food 41
Sri Lanka: trees 127–9
state: in conflicts 141; enabling 160;
    and entitlement concept 36–7,
    64–86; and hunger 148–9, 159–60;
    land reform 67; role in mitigating
    famine 12, 14–15
statistics: hunger 26–30; since 1970s
    147–8
Stonich, S. 73–4
street vendors: food 79–82
structural adjustment policies (SAPs)
    43–6, 167; Zimbabwe 75
Sudan: cattle 137; siege 138
sugar production: Brazil 32–3;
    Caribbean 37–8
Synder, M. 98–9

Taiwan: land ownership 67
Tansey, G. 52–3, 58
technology: and decline in mortality
    rates 11–12; food 40; and gender 98;
    global cool chains 49; lack of in
    developing countries 59;
    relationship with hunger 150; role of
    state 160; transport 38
third world countries see developing
    countries
Thubron, C. 120–1
Tinker, I. 96
trading, international: future changes
    149–52
transnational corporations (TNCs) 150;
    power of 58–9; role in global food
    system 49–54
transport, food 37–8; technology 38
trees: jak and breadfruit 127–9; as raw
    material 130; role of 126–30

UK see Britain

undernutrition 25–6; causes 4; in Sri
  Lanka 127–9
UNICEF 19–20, 24
United Nations (UN): Afghanistan
  142–3; breastfeeding debate 24;
  international aid 152–3; refugees
  139, 140; *see also* Food and
  Agricultural Organisation
United Nations Decade of Women
  (1975–85) 90–1
United Nations Development Program
  (UNDP) 89, 90, 105
urban areas: development strategies
  70–83; entitlement 6; food
  production 82–3; food supplies
  78–83; India 112–15
Uruguay Round 46, 149
USA: Alliance for Progress 73; cereals
  151, 152; dominance in agricultural
  trade 46; entitlement 6, 41; infant
  mortality rates 25, 31; in integrated
  market 49; low birthweights 25
Uvin, P. 17–18, 19, 25, 30, 156, 161,
  166

Venezuela: food riots 45
Vietnam: shrimp and chicken salad
  recipe 1
villagisation: Mozambique 138
vitamin A deficiency 20
vulnerability, spaces of 4; Great Irish
  Famine 8–10

Walton, J. 42, 44–5
war *see* conflict
Warnock, J.W. 2, 33
water: access to 24–5, 148; semi-arid
  areas 115–16
Watkins, K. 137, 149, 153, 154, 163–4

Watts, M.J. 4
weight: as indication of undernutrition
  25, 105
wheat: as staple food 38–40
widows: Indian 108–9
Wilson, K. 136
women: anaemia 20–1; delayed
  menarche 25; development policies
  167; as economic actors 91–9; fish
  farming 78; food entitlement 88–109;
  and hunger 33–4; malnourished
  21–2; missing 89; and population
  growth 3; refugee 140
Women's Feature Service 94–5
women's groups 106
working classes: in nineteenth-century
  England 8–9
World Alliance for Breastfeeding
  Action (WABA) 24
World Bank (WB) 43, 75, 161, 162–3,
  164; development funds 154; and
  gender 91; industrial development 71
World Food Programme: aid to North
  Korea 4
World Health Organisation (WHO) 19;
  iodine deficiency 20
World Summit for Children (1990) 20,
  155–6
World Trade Organization (WTO) 46,
  154
Worsley, T. 52–3, 58

Young, E.M. 10

Zambia: food riots 45; women
  agricultural workers 96–7
Zimbabwe 44; food security 75; urban
  food supplies 79–83
Zwi, A. 133, 135, 137